---

186 Dimensions, Embeddings, and Attractors

Dimensions, Embeddings, and Attractors

JAMES C. ROBINSON
University of Warwick

CAMBRIDGE
UNIVERSITY PRESS

University Printing House, Cambridge CB2 8BS, United Kingdom

One Liberty Plaza, 20th Floor, New York, NY 10006, USA

477 Williamstown Road, Port Melbourne, VIC 3207, Australia

314-321, 3rd Floor, Plot 3, Splendor Forum, Jasola District Centre, New Delhi - 110025, India

103 Penang Road, #05-06/07, Visioncrest Commercial, Singapore 238467

Cambridge University Press is part of the University of Cambridge.

It furthers the University's mission by disseminating knowledge in the pursuit of education, learning and research at the highest international levels of excellence.

www.cambridge.org
Information on this title: www.cambridge.org/9780521898058

First published 2011

A catalogue record for this publication is available from the British Library

Library of Congress Cataloging in Publication data
Robinson, James C. (James Cooper), 1969–
Dimensions, Embeddings, and Attractors / James C. Robinson.
p. cm. – (Cambridge Tracts in Mathematics ; 186)
Includes bibliographical references and index.
ISBN 978-0-521-89805-8 (hardback)
1. Dimension theory (Topology) 2. Attractors (Mathematics)
3. Topological imbeddings. I. Title. II. Series.
QA611.3.R63 2011
515′.39 – dc22 2010042726

ISBN 978-0-521-89805-8 Hardback

To my family: Tania, Joseph, & Kate.

Contents

Preface

The main purpose of this book is to bring together a number of results concerning the embedding of 'finite-dimensional' compact sets into Euclidean spaces, where an 'embedding' of a metric space (X, ϱ) into $\mathbb{R}^n$ is to be understood as a homeomorphism from X onto its image. A secondary aim is to present, alongside such 'abstract' embedding theorems, more concrete embedding results for the finite-dimensional attractors that have been shown to exist in many infinite-dimensional dynamical systems.

In addition to its summary of embedding results, the book also gives a unified survey of four major definitions of dimension (Lebesgue covering dimension, Hausdorff dimension, upper box-counting dimension, and Assouad dimension). In particular, it provides a more sustained exposition of the properties of the box-counting dimension than can be found elsewhere; indeed, the abstract results for sets with finite box-counting dimension are those that are taken further in the second part of the book, which treats finite-dimensional attractors.

While the various measures of dimension discussed here find a natural application in the theory of fractals, this is not a book about fractals. An example to which we will return continually is an orthogonal sequence in an infinite-dimensional Hilbert space, which is very far from being a 'fractal'. In particular, this class of examples can be used to show the sharpness of three of the embedding theorems that are proved here.

My models have been the classic text of Hurewicz & Wallman (1941) on the topological dimension, and of course Falconer's elegant 1985 tract which concentrates on the Hausdorff dimension (and Hausdorff measure). It is a pleasure to acknowledge formally my indebtedness to Hunt & Kaloshin's 1999 paper 'Regularity of embeddings of infinite-dimensional fractal sets into finite-dimensional spaces'. It has had a major influence on my own research over the last ten years, and one could view this book as an extended exploration of the ramifications of the approach that they adopted there.

My interest in abstract embedding results is related to the question of whether one can reproduce the dynamics on a finite-dimensional attractor using a finite-dimensional system of ordinary differential equations (see Chapter 10 of Eden, Foias, Nicolaenko, & Temam (1994), or Chapter 16 of Robinson (2001), for example). However, there are still only partial results in this direction, so this potential application is not treated here; for an up-to-date discussion see the paper by Pinto de Moura, Robinson, & Sánchez-Gabites (2010).

I started writing this book while I was a Royal Society University Research Fellow, and many of the results here derive from work done during that time. I am currently supported by an EPSRC Leadership Fellowship, Grant EP/G007470/1. I am extremely grateful to both the Royal Society and to the EPSRC for their support.

I would like to thank Alexandre Carvalho, Peter Friz, Igor Kukavica, José Langa, Eric Olson, Eleonora Pinto de Moura, and Alejandro Vidal López, all of whom have had a hand in material that is presented here. In particular, Eleonora was working on closely-related problems for her doctoral thesis during most of the time that I was writing this book, and our frequent discussions have shaped much of the content and my approach to the material. I had comments on a draft version of the manuscript from Witold Sadowski, Jaime Sánchez-Gabites, and Nicholas Sharples: I am extremely grateful for their helpful and perceptive comments. David Tranah, Clare Dennison, and Emma Walker at Cambridge University Press have been most patient as one deadline after another was missed and extended; that one was finally met (nearly) is due in large part to a kind invitation from Marco Sammartino to Palermo, where I gave a series of lectures on some of the material in this book in November 2009.

Many thanks to my parents and to my mother-in-law; in addition to all their other support, their many days with the children have made this work possible. Finally, of course, thanks to Tania, my wife, and our children Joseph and Kate, who make it all worthwhile; this book is dedicated to them.

Introduction

Part I of this book treats four different definitions of dimension, and investigates what being 'finite dimensional' implies in terms of embeddings into Euclidean spaces for each of these definitions.

Whitney (1936) showed that any abstract n-dimensional C^r manifold is C^r-homeomorphic to an analytic submanifold in $\mathbb{R}^{2n+1}$. This book treats embeddings for much more general sets that need not have such a smooth structure; one might say 'fractals', but we will not be concerned with the fractal nature of these sets (whatever one takes that to mean).

We will consider four major definitions of dimension:

(i) The (Lebesgue) covering dimension $\dim(X)$, based on the maximum number of simultaneously intersecting sets in refinements of open covers of X (Chapter 1). This definition is topologically invariant, and is primarily used in the classical and abstract 'Dimension Theory', elegantly developed in Hurewicz & Wallman's 1941 text, and subsequently by Engelking (1978), who updates and extends their treatment.

(ii) The Hausdorff dimension $d_{\mathrm{H}}(X)$, the value of d where the 'd-dimensional Hausdorff measure' of X switches from ∞ to zero (Chapter 2). Hausdorff measures (and hence the Hausdorff dimension) play a large role in geometric measure theory (Federer, 1969), and in the theory of dynamical systems (see Pesin (1997)); the standard reference is Falconer's 1985 tract, and subsequent volumes (Falconer, 1990, 1997).

(iii) The (upper) box-counting dimension $d_{\mathrm{B}}(X)$, essentially the scaling as $\epsilon \to 0$ of $N(X, \epsilon)$, the number of ϵ-balls required to cover X, i.e. $N(X, \epsilon) \sim \epsilon^{-d_{\mathrm{B}}(X)}$ (Chapter 3). This dimension has mainly found application in the field of dynamical systems, see for example Falconer (1990), Eden *et al.* (1994), C. Robinson (1995), and Robinson (2001).

(iv) The Assouad dimension $d_A(X)$, a 'uniform localised' version of the box-counting dimension: if $B(x, \rho)$ denotes the ball of radius ρ centred at $x \in X$, then $N(X \cap B(x, \rho), r) \sim (\rho/r)^{d_A(X)}$ for every $x \in X$ and every $0 < r < \rho$ (Chapter 9). This definition appears unfamiliar outside the area of metric spaces and most results are confined to research papers (e.g. Assouad (1983), Luukkainen (1998), Olson (2002); but see also Heinonen (2001, 2003)).

For any compact metric space (X, ϱ) we will see that

$$\dim(X) \leq d_H(X) \leq d_B(X) \leq d_A(X),$$

and there are examples showing that each of these inequalities can be strict. We will check that each definition satisfies the natural properties of a dimension: monotonicity ($X \subseteq Y$ implies that $d(X) \leq d(Y)$); stability under finite unions ($d(X \cup Y) = \max(d(X), d(Y))$); and the dimension of $\mathbb{R}^n$ is n (a consistent way to interpret this so that it makes sense for all the definitions above is that $d(K) = n$ if K is a compact subset of $\mathbb{R}^n$ that contains an open set). We will also consider how each definition behaves for product sets.

Our main concern will be with the embedding results that are available for each class of 'finite-dimensional' set. The embedding result for sets with finite covering dimension, due to Menger (1926) and Nöbeling (1931) (given as Theorem 1.12 here), is in a class of its own. The result guarantees that when $\dim(X) \leq d$, a generic set of continuous maps from a compact metric space (X, ϱ) into $\mathbb{R}^{2d+1}$ are embeddings.

The results for sets with finite Hausdorff, upper box-counting, and Assouad dimension are of a different cast. They are expressed in terms of 'prevalence' (a version of 'almost every' that is applicable to subsets of infinite-dimensional spaces, introduced independently by Christensen (1973) and Hunt, Sauer, & Yorke (1992), and the subject of Chapter 5), and treat compact subsets of Hilbert and Banach spaces. Using techniques introduced by Hunt & Kaloshin (1999), we show that a 'prevalent' set of continuous linear maps $L : \mathscr{B} \to \mathbb{R}^k$ provide embeddings of X when $d(X - X) < k$, where

$$X - X = \{x_1 - x_2 : x_1, x_2 \in X\}$$

and d is one of the above three dimensions (see Figure 1). Note that if one wishes to show that a linear map provides an embedding, i.e. that $Lx = Ly$ implies that $x = y$, this is equivalent to showing that $Lz = 0$ implies that $z = 0$ for $z \in X - X$. This is why the natural condition for such results is one on the 'difference' set $X - X$; but while $d_B(X - X) \leq 2d_B(X)$, there are examples of

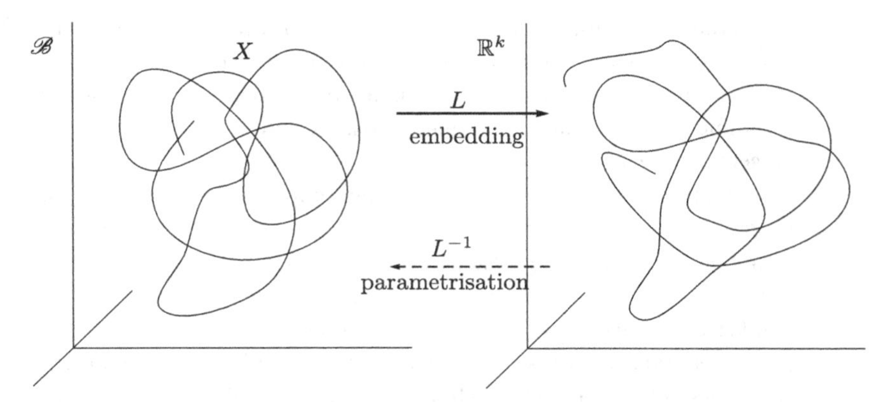

Figure 1 The linear map $L : \mathscr{B} \to \mathbb{R}^k$ embeds X into $\mathbb{R}^k$. The inverse mapping L^{-1} provides a parametrisation of X using k parameters.

sets for which $d_{\rm H}(X) = 0$ but $d_{\rm H}(X - X) = \infty$ (and similarly for the Assouad dimension).

Where the embedding results for these three dimensions differ from one another is in the smoothness of the parametrisation of X provided by L^{-1}. In the Hausdorff case this inverse can only be guaranteed to be continuous (Chapter 6); in the upper box-counting case it will be Hölder (Chapter 8); and in the Assouad case it will be Lipschitz to within logarithmic corrections (Chapter 9). Simple examples of orthogonal sequences in ℓ^2 (or related examples in c_0, the space of sequences that tend to zero) show that the results we give cannot be improved when the embedding map L is linear.

Chapter 4 presents an embedding result for subsets X of $\mathbb{R}^N$ with box-counting dimension $d < (N - 1)/2$. The ideas here form the basis of the results for subsets of Hilbert and Banach spaces that follow, and justify the development of the theory of prevalence in Chapter 5 and the definition of various 'thickness exponents' (the thickness exponent itself, the Lipschitz deviation, and the dual thickness) in Chapter 7.

Part II discusses the attractors that arise in certain infinite-dimensional dynamical systems, and the implications of the results of Part I for this class of finite-dimensional sets. In particular, the embedding result for sets with finite box-counting dimension is used toward a proof of an infinite-dimensional version of the Takens time-delay embedding theorem (Chapter 14) and it is shown that a finite-dimensional set of real analytic functions can be parametrised using a finite number of point values (Chapter 15).

Chapter 10 gives a very cursory summary of some elements of the theory of Sobolev spaces and fractional power spaces of linear operators, which are required in order to discuss the applications to partial differential equations. It is shown how the solutions of an abstract semilinear parabolic equation, and of the two-dimensional Navier–Stokes equations, can be used to generate an infinite-dimensional dynamical system whose evolution is described by a nonlinear semigroup.

The global attractor of such a nonlinear semigroup is a compact invariant set that attracts all bounded subsets of the phase space. A sharp condition guaranteeing the existence this global attractor is given in Chapter 11, and it is shown that such an object exists for the semilinear parabolic equation and the Navier–Stokes equations that were treated in the previous chapter.

Chapter 12 provides a method for bounding the upper box-counting dimension of attractors in Banach spaces. While there are powerful techniques available for attractors in Hilbert spaces, these are already presented in a number of other texts, and outlining the more general Banach space technique is more in keeping with the overall approach of this book (the Hilbert space method is covered here in an extended series of exercises). In particular, we show that any attractor of the abstract semilinear parabolic equation introduced in Chapter 10 will be finite-dimensional.

Before proving the final two 'concrete' embedding theorems in Chapters 14 and 15, Chapter 13 provides two results that guarantee that an attractor has zero 'thickness': we show first that if the attractor consists of smooth functions then its thickness exponent is zero, and then that the attractors of a wide variety of models (which can be written in the abstract semilinear parabolic form) have zero Lipschitz deviation. This, in part, answers a conjecture of Ott, Hunt, & Kaloshin (2006).

Most of the chapters end with a number of exercises. Many of these carry forward portions of the argument that would break the flow of the main text, or discuss related approaches. Full solutions of the exercises are given at the end of the book.

All Hilbert and Banach spaces are real, throughout.

PART I

Finite-dimensional sets

1
Lebesgue covering dimension

There are a number of definitions of dimension that are invariant under homeomorphisms, i.e. that are topological invariants – in particular, the large and small inductive dimensions, and the Lebesgue covering dimension. Although different a priori, the large inductive dimension and the Lebesgue covering dimension are equal in any metric space (Katětov, 1952; Morita, 1954; Chapter 4 of Engelking, 1978), and all three definitions coincide for separable metric spaces (Proposition III.5 A and Theorem V.8 in Hurewicz & Wallman (1941)). A beautiful exposition of the theory of 'topological dimension' is given in the classic text by Hurewicz & Wallman (1941), which treats separable spaces throughout and makes much capital out of the equivalence of these definitions. Chapter 1 of Engelking (1978) recapitulates these results, while the rest of his book discusses dimension theory in more general spaces in some detail.

This chapter concentrates on one of these definitions, the Lebesgue covering dimension, which we will denote by $\dim(X)$, and refer to simply as the covering dimension. Among the three definitions mentioned above, it is the covering dimension that is most suitable for proving an embedding result: we will show in Theorem 1.12, the central result of this chapter, that if $\dim(X) \leq n$ then a generic set of continuous maps from X into $\mathbb{R}^{2n+1}$ are homeomorphisms, i.e. provide an embedding of X into $\mathbb{R}^{2n+1}$.

There is, unsurprisingly, a topological flavour to the arguments involved here, and consequently they are very different from those in the rest of this book. However, any survey of embedding results for finite-dimensional sets would be incomplete without including the 'fundamental' embedding theorem that is available for sets with finite covering dimension.

1.1 Covering dimension

Let (X, ϱ) be a metric space, and A a subset[1] of X. A *covering* of $A \subseteq X$ is a finite collection $\{U_j\}_{j=1}^r$ of open subsets of X such that

$$A \subseteq \bigcup_{j=1}^{r} U_j.$$

The *order* of a covering is the largest integer n such that there are $n+1$ members of the covering that have a nonempty intersection. A covering β is a *refinement* of a covering α if every member of β is contained in some member of α.

Definition 1.1 A set $A \subseteq X$ has $\dim(A) \leq n$ if every covering has a refinement of order $\leq n$. A set A has $\dim(A) = n$ if $\dim(A) \leq n$ but it is not true that $\dim(A) \leq n - 1$.

Clearly dim is a topological invariant. We now prove some elementary properties of the covering dimension, following Munkres (2000) and Edgar (2008).

Proposition 1.2 *Let $B \subseteq A \subseteq X$, with B closed. If $\dim(A) = n$ then $\dim(B) \leq n$.*

Proof Let α be a covering of B by open subsets $\{U_j\}$ of X. Cover A by the sets $\{U_j\}$, along with the open set $X \setminus B$. Let β be a refinement of this covering that has order at most n. Then the collection

$$\beta' := \{U \in \beta : U \cap B \neq \emptyset\}$$

is a refinement of α that covers B and has order at most n. □

The assumption that B is closed makes the proof significantly simpler, but the result remains true for an arbitrary subset of A, see Theorem 3.2.13 in Edgar (2008), or Theorem III.1 in Hurewicz & Wallman (1941). However, the following 'sum theorem' is not true unless one of the spaces is closed: in fact, $\dim(X) = n$ if and only if X can be written as the union of $n + 1$ subsets all of which have dimension zero (see Theorem III.3 in Hurewicz & Wallman (1941)).

[1] In the context of metric spaces it is somewhat artificial to make the definition in this form, since (A, ϱ) is a metric space in its own right. But our main focus in what follows will be on subsets of Hilbert and Banach spaces, where the underlying linear structure of the ambient space will be significant.

Proposition 1.3 *Let $X = X_1 \cup X_2$, where X_1 and X_2 are closed subspaces of X with $\dim(X_1) \leq n$ and $\dim(X_2) \leq n$. Then $\dim(X) \leq n$.*

Of course, it follows that if $X = X_1 \cup \cdots \cup X_k$, each X_j is closed and $\dim(X_j) \leq n$ for every $j = 1, \ldots, k$ then $\dim(X) \leq n$. In fact one can extend this to countable unions of closed sets, see Theorem III.2 in Hurewicz & Wallman (1941) (and Theorem 3.2.11 in Edgar (2008) for the case $n = 1$).

Proof We will say that an open covering α of X has order at most n at points of Y if every point in Y lies in no more than $n + 1$ elements of α.

First we show that any open covering α of X has a refinement that has order at most n at points of X_1. Any such covering of X provides a covering of X_1, which has a refinement β' that has order at most n. For every $V \in \beta'$, there exists an element $U_V \in \alpha$ such that $V \subset U_V$. Then

$$\beta = \{U_V : \; V \in \beta'\} \cup \{U \setminus X_1 : \; U \in \alpha\}$$

is the required refinement of α. We can repeat this argument starting with the covering β of X, and obtain a covering γ that refines β and has order at most n at points of X_2.

We now define a further covering of X, which will turn out to be a refinement of α of order at most n. As a first step in our construction, define a map $f : \gamma \to \beta$ by choosing, for each $G \in \gamma$, an $f(G) \in \beta$ such that $G \subset f(G)$ (this is possible since γ refines β). Now for each $B \in \beta$, let

$$d(B) = \{G \in \gamma : \; f(G) = B\},$$

and let δ be the union of all the sets $d(B)$ (over $B \in \beta$).

Now, δ is a refinement of α, since $d(B) \subset B$ for every $B \in \beta$, and β is a refinement of α. Also, δ still covers X since γ covers X and every $G \in \gamma$ is contained in some $B \in \beta$ (as γ refines β). All that remains is to show that δ has order at most n.

Suppose that $x \in X$ with $x \in d(B_1) \cap \cdots \cap d(B_k)$, with all the $d(B_k)$ distinct (thus $B_1, \ldots, B_k$ are distinct). It follows that for each $j = 1, \ldots, k$, $x \in G_j$ where $f(G_j) = B_j$; since $B_1, \ldots, B_k$ are distinct, so are $G_1, \ldots, G_k$. Thus

$$x \in G_1 \cap \cdots \cap G_k \;\subset\; d(B_1) \cap \cdots \cap d(B_k) \;\subset\; B_1 \cap \cdots \cap B_k.$$

If $x \in X_1$ then $k \leq n + 1$ because β has order at most n at points of X_1; and if $x \in X_2$ then $k \leq n + 1$ because γ has order at most n at points of X_2. □

We do not prove a result on the covering dimension of products here, although it is the case that $\dim(X \times Y) \leq \dim(X) + \dim(Y)$ (Theorem III.4 in

Hurewicz & Wallman (1941)): this can be proved as a corollary of a characterisation of the covering dimension in terms of the upper box-counting dimension, see Exercise 3.4.

1.2 The covering dimension of I_n

It is by no means trivial to show that the covering dimension of $\mathbb{R}^n$ is n. Note that it suffices to show that $\dim(I_n) = n$, where $I_n = [-\frac{1}{2}, \frac{1}{2}]^n$ denotes the unit cube in $\mathbb{R}^n$, since as remarked after Proposition 1.3, the covering dimension is in fact stable under countable unions of closed sets.

We refer to Theorem 50.6 in Munkres (2000) for a direct proof of the upper bound on $\dim(I_n)$ (see also Exercise 1.2 for compact subsets of $\mathbb{R}^2$). One can also deduce the upper bound from the general fact that the covering dimension is bounded by the Hausdorff dimension (Theorem 2.11); it is very simple to show that the Hausdorff dimension of a subset of $\mathbb{R}^n$ is bounded by n (Proposition 2.8(iii)).

While the proof of the upper bound is more notationally awkward than technically difficult, the proof of the lower bound involves the powerful Brouwer Fixed Point Theorem (see IV (C) of Hurewicz & Wallman (1941) for a proof).

Theorem 1.4 *Any continuous map $f : I_n \to I_n$ has a fixed point, i.e. there exists an $x_0 \in I_n$ such that $f(x_0) = x_0$.*

We give a proof of the lower bound (essentially the 'Lebesgue Covering Theorem') adapted from Hurewicz & Wallman's book, for the two-dimensional unit cube $I_2 = [-\frac{1}{2}, \frac{1}{2}]^2$. The general result (for I_n) is not significantly more involved, but the argument can be somewhat simplified in this case without losing its essential flavour. (An alternative proof of a similar two-dimensional result is given as Theorem 3.3.4 in Edgar (2008).) Before the proof we introduce some notation.

Given a set $U \subset (X, \varrho)$ we define the diameter of U, written $|U|$, as

$$|U| = \mathrm{diam}(U) = \sup_{u_1, u_2 \in U} \varrho(u_1, u_2).$$

(We only use the notation $\mathrm{diam}(U)$ when $|U|$ would be ambiguous.) The *mesh size* of a covering of A is the largest of the diameters of the elements of the covering.

For two sets $A, B \subset X$ we write

$$\mathrm{dist}(A, B) = \sup_{a \in A} \inf_{b \in B} \varrho(a, b)$$

for the Hausdorff semidistance between A and B. Note that if B is closed then $\mathrm{dist}(A, B) = 0$ implies that $A \subseteq B$.

Theorem 1.5 *Let* $I_2 = [-\frac{1}{2}, \frac{1}{2}]^2 \subset \mathbb{R}^2$. *Then* $\dim(I_2) \geq 2$.

Proof We want to show that any covering α of I_2 with sufficiently small mesh size contains at least three sets with nonempty intersection. To this end, take a covering α with mesh size < 1 so that no element of the covering contains points of opposite faces.

The first step is to construct a refinement $\tilde{\alpha}$ of α consisting of closed, rather than open, sets. To do this, observe that every $x \in I_2$ is contained in some $U_x \in \alpha$, and we can find an open set V_x such that $x \in V_x \subset \bar{V}_x \subset U_x$. Since I_2 is compact and $\{V_x : x \in I_2\}$ is an open cover of I_2, there is a finite subcover $\{V_{x_j}\}$. We take $\tilde{\alpha}$ to be the collection of all the closed sets $\{\bar{V}_{x_j}\}$. By construction this is a refinement of α consisting of closed sets.

We now show that $\tilde{\alpha}$ contains at least three sets with nonempty intersection, from which it is immediate (since $\tilde{\alpha}$ is a refinement of α) that α contains at least three sets with nonempty intersection.

Let Γ_1 denote the side of I_2 with $x = -\frac{1}{2}$, Γ_1' the side with $x = \frac{1}{2}$, Γ_2 the side with $y = -\frac{1}{2}$, and Γ_2' the side with $y = \frac{1}{2}$. Let L_1 denote the union of those elements of $\tilde{\alpha}$ that intersect Γ_1; L_2 the union of those elements of $\tilde{\alpha}$ that are not in L_1 and intersect Γ_2; and let L_3 be the union of all the other elements of $\tilde{\alpha}$ (those that intersect neither Γ_1 nor Γ_2). See Figure 1.1(a).

If we define $K_1 = L_1 \cap L_3$ then K_1 separates Γ_1 and Γ_1' in I_2, i.e. there exist open sets U_1 and U_1' such

$$I_2 \setminus K_1 = U_1 \cup U_1', \qquad U_1 \cap U_1' = \emptyset$$

and $\Gamma_1 \subset U_1$, $\Gamma_1' \subset U_1'$. The set $K_2' = L_1 \cap L_2 \cap L_3$ separates $\Gamma_2 \cap K_1$ from $\Gamma_2' \cap K_1$ in K_1. One can then find a new closed set K_2, with $K_2 \cap K_1 \subseteq K_2'$, that separates Γ_2 and Γ_2' in I_2, i.e. such that there exist open sets U_2 and U_2' such that

$$I_2 \setminus K_2 = U_2 \cup U_2', \qquad U_2 \cap U_2' = \emptyset$$

and $\Gamma_2 \subset U_2$, $\Gamma_2' \subset U_2'$. These constructions are illustrated in Figure 1.1(b). (If the 'proof by diagram' of this last step is unconvincing, see IV.3 A) in Hurewicz & Wallman (1941), or Exercise 1.3.)

Now for each $x \in I_2$, let $v(x)$ be the 2-vector with components

$$v_i(x) = \begin{cases} \mathrm{dist}(x, K_i) & x \in U_i, \\ 0 & x \in K_i, \\ -\mathrm{dist}(x, K_i) & x \in U_i', \end{cases}$$

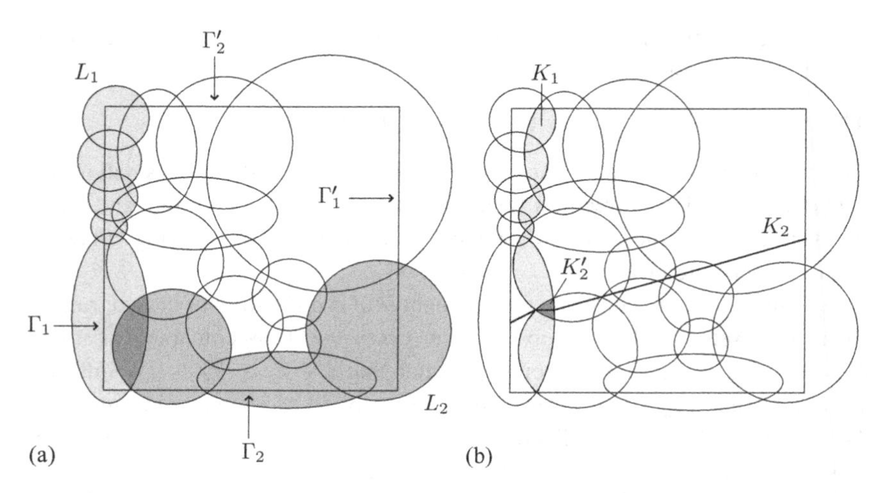

Figure 1.1 (a) A covering of I_2, divided into sets L_1 (lightly shaded), L_2 (more heavily shaded), and L_3 (not shaded). (b) K_1 (lightly shaded) separates Γ_1 and Γ_1' in I_2; K_2' (a subset of K_1, shaded more heavily) separates $K_1 \cap \Gamma_2$ and $K_1 \cap \Gamma_2'$ in K_1; K_2 (the dark line) separates Γ_2 and Γ_2' in I_2, with $K_2 \cap K_1 \subseteq K_2'$.

and set $f(x) = x + v(x)$; note that $f(x) \in I_2$, and that f is continuous. It follows from the Brouwer Fixed Point Theorem (Theorem 1.4) that f has a fixed point, i.e. there exists an $x_0 \in I_2$ such that $f(x_0) = x_0$. In particular, this implies that $\text{dist}(x_0, K_1) = \text{dist}(x_0, K_2) = 0$, i.e. that $K_1 \cap K_2 \subset K_2' = L_1 \cap L_2 \cap L_3$ is nonempty. Since each of the original elements of $\tilde{\alpha}$ is contained in only one of the L_js, there are three elements of $\tilde{\alpha}$ that contain a common point. $\square$

1.3 Embedding sets with finite covering dimension

We now prove the fundamental embedding result that any space with covering dimension n can be topologically embedded into $\mathbb{R}^{2n+1}$; note that this characterises sets of finite covering dimension as homeomorphic images of subsets of finite-dimensional Euclidean spaces. The embedding result in the compact case (which we treat here) is due to Menger (1926) and Nöbeling (1931); we follow the presentation of Hurewicz & Wallman (1941, Theorem V.2) and Munkres (2000, Theorem 50.5). A similar result is possible in the general (non compact) case, see Theorem V.3 in Hurewicz & Wallman (1941).

The proof uses the Baire Category Theorem, which we state here for convenience. For a proof see Munkres (2000, Theorem 48.2), for example.

Theorem 1.6 *Let (X, ϱ) be a complete metric space, and $\{X_j\}_{j=1}^{\infty}$ a countable collection of open and dense subsets of X. Then*

$$\bigcap_{j=1}^{\infty} X_j$$

is a dense subset of X.

We begin with a useful characterisation of the covering dimension of compact sets.

Lemma 1.7 *A compact set $A \subseteq (X, \varrho)$ has $\dim(A) \le n$ if and only if it has coverings of arbitrarily small mesh size and order $\le n$.*

Proof First we show that if α is a covering of A then there is an $\eta > 0$ such that every subset of A with diameter less than η is entirely contained in some member of α (the largest such η is called the 'Lebesgue number' of the covering α). If not, there exists a sequence $\{A_j\}_{j=1}^{\infty}$ of subsets of A with diameters tending to zero not wholly contained in any member of α. Choose $x_j \in A_j$; since A is compact, there exists a subsequence (which we relabel) such that $x_j \to x^*$. Of course, $x^* \in U \in \alpha$. But since U is open, $B(x^*, \delta) \subset U$ for some $\delta > 0$, from which it follows that $A_j \subset U \in \alpha$ for all j sufficiently large, contradicting our initial assumption.

So now take an initial covering α of A. By assumption there exists a covering β of A of mesh size $< \eta$ and of order $\le n$; we have just shown that each element of this covering β lies entirely within an element of α. It follows that β is a refinement of α of order $\le n$, and so $\dim(A) \le n$.

Now suppose that $\dim(A) \le n$. Consider the collection of all open balls in A of radius $\epsilon/2$. Since A is compact, there is a covering of A by a finite collection of these balls. It follows from the fact $\dim(A) \le n$ that there is a refinement of this covering (still consisting of sets whose diameter is no larger than ϵ) of order $\le n$. □

We say that a continuous map $g : X \to \mathbb{R}^k$ is an ϵ-mapping if

$$\mathrm{diam}[g^{-1}(x)] < \epsilon \qquad \text{for all} \qquad x \in g(X).$$

We will show that for each $n \in \mathbb{N}$, the set of all $1/n$-mappings is open and dense, and our embedding result will then follow using the Baire Category Theorem (Theorem 1.6) and the following simple lemma.

Lemma 1.8 *If (X, ϱ) is compact then g is a homeomorphism of X into $\mathbb{R}^k$ if and only if g is a $1/n$-mapping for each $n \in \mathbb{N}$.*

Proof If g is a $1/n$-mapping for each $n \in \mathbb{N}$ then $\text{diam}[g^{-1}(x)] = 0$ for every $x \in g(X)$, i.e. $g^{-1}(x)$ consists of a single point, so that g is one-to-one. But a one-to-one continuous mapping of a compact set is a homeomorphism, see Exercise 1.4. The converse is clear. □

Lemma 1.9 *Let (X, ϱ) be compact. Then for each $\epsilon > 0$ the set $\mathscr{F}_\epsilon$ of all ϵ-mappings is open in $C(X, \mathbb{R}^k)$.*

Proof Suppose that $g \in C(X, \mathbb{R}^k)$ is an ϵ-mapping. Since X is compact so is $X \times X$, and since

$$\{(x, x') \in X \times X \text{ with } \varrho(x, x') \geq \epsilon\}$$

is a closed subset of $X \times X$, it too is compact. It follows that

$$\eta = \inf\{|g(x) - g(x')| : x, x' \in X \text{ with } \varrho(x, x') \geq \epsilon\} > 0;$$

if η were zero then g could not be an ϵ-mapping. If f is any mapping with $\varrho(f, g) < \eta/2$ and $f(x) = f(x')$ it follows that $|g(x) - g(x')| < \eta$, and hence that $\varrho(x, x') < \epsilon$, i.e. f is also an ϵ-mapping. □

The density of $\mathcal{F}_\epsilon$ is much more delicate, and requires the following geometric result, for which we follow the presentation in Munkres (2000). Given a collection $\{x_1, \ldots, x_k\}$ of two or more points in $\mathbb{R}^N$, the affine space generated by $\{x_1, \ldots, x_k\}$, $A(x_1, \ldots, x_k)$ is the collection of all points of the form

$$\sum_{j=1}^{k} a_j x_j \qquad \text{with} \qquad \sum_{j=1}^{k} a_j = 1.$$

It is easy to check that this is the same as the affine space through x_1 spanned by $\{x_j - x_1\}_{j=2}^k$, i.e. all points of the form

$$x_1 + \sum_{j=2}^{k} c_j (x_j - x_1)$$

for all $c_j \in \mathbb{R}$ (one could also form the same space by taking any x_i and considering all points of the form $x_i + \sum_{j \neq i} c_j (x_j - x_i)$). We say that the points $\{x_1, \ldots, x_n\}$ in $\mathbb{R}^N$ are in general position in $\mathbb{R}^N$ if no x_j lies in the affine space generated by any subcollection of the $\{x_i\}$ consisting of $\leq N$ elements that does not contain x_j.

An equivalent and more elegant definition makes use of the following concept. We say that a set $\{y_1, \ldots, y_k\}$ of k points in $\mathbb{R}^N$ are *geometrically independent* if

$$\sum_{j=1}^{k} a_j y_j = 0 \qquad \text{and} \qquad \sum_{j=1}^{k} a_j = 0,$$

then $a_j = 0$ for $j = 1, \ldots, k$. A set of points in $\mathbb{R}^N$ are in general position if any k of these points, $k \le N + 1$, are geometrically independent. Note that if the points $\{y_1, \ldots, y_k\}$ are geometrically independent then so are the points $\{y_1 + z, \ldots, y_k + z\}$ for any $z \in \mathbb{R}^N$.

(For later use we note that the simplex spanned by $\{x_1, \ldots, x_k\}$ is the convex hull of $\{x_1, \ldots, x_k\}$; or equivalently the affine combinations $\sum_{i=1}^{k} a_i x_i$ with $\sum_{i=1}^{k} a_i = 1$ and $a_i \ge 0$ for every $i = 1, \ldots, k$. If the $\{x_j\}_{j=1}^{k}$ are geometrically independent then the dimension of the simplex is (by definition) $k - 1$. A polyhedron is a finite union of simplices in some $\mathbb{R}^N$; its dimension is (by definition) the maximum of the dimension of these simplices.)

We now show that near any collection of points in $\mathbb{R}^N$, there is a set of points that are in general position.

Lemma 1.10 *Given points $x_1, \ldots, x_n \in \mathbb{R}^N$ and $\delta > 0$, there exist points $y_1, \ldots, y_n \in \mathbb{R}^N$ such that $|x_i - y_i| < \delta$ and the $\{y_1, \ldots, y_n\}$ are in general position in $\mathbb{R}^N$.*

Proof We prove this by induction. Suppose that we have a collection of $k - 1$ points, $\{y_1, \ldots, y_{k-1}\}$, in general position in $\mathbb{R}^N$. There are a finite number of subcollections of the y_is consisting of $\le N$ elements, each of which generates an affine subspace of dimension $\le N - 1$. The measure of each of these subspaces is zero; and so is the measure of their union S. So there certainly exists a $y_k \in B(x_k, \delta) \setminus S$. The set $\{y_1, \ldots, y_k\}$ is in general position: indeed, if we choose $\le N + 1$ of these y_j, either none of these is y_k in which case the induction hypothesis guarantees that they are geometrically independent, or one of them is y_k which we have just constructed to ensure geometric independence. □

Proposition 1.11 *Let (X, ϱ) be a compact metric space of dimension $\le n$. Then for each $\epsilon > 0$, $\mathscr{F}_\epsilon$ is dense in $C(X, \mathbb{R}^{2n+1})$.*

Proof Take $f \in C(X, \mathbb{R}^{2n+1})$ and $\eta > 0$. We will construct a $g \in \mathscr{F}_\epsilon$ such that $\varrho(f, g) < \eta$.

Since X is compact, f is uniformly continuous and so there exists a $\delta < \epsilon$ such that

$$\varrho(x, x') < \delta \qquad \Rightarrow \qquad \varrho(f(x), f(x')) < \eta/2. \tag{1.1}$$

Since X is compact and $\dim(X) \le n$, there exists a covering $\{U_j\}_{j=1}^r$ of X of order $\le n$ such that $\operatorname{diam}(U_j) < \delta$ for all j. It follows from (1.1) that

$$\operatorname{diam}(f(U_j)) < \eta/2 \tag{1.2}$$

for each j.

Now use Lemma 1.10 to select points $\{p_j\}_{j=1}^r$ in $\mathbb{R}^{2n+1}$ such that

$$\operatorname{dist}(p_j, f(U_j)) < \eta/2 \tag{1.3}$$

and the $\{p_1, \ldots, p_r\}$ are in general position in $\mathbb{R}^{2n+1}$.

For each point $x \in X$ and $1 \le i \le r$ define

$$w_i(x) = \operatorname{dist}(x, X \setminus U_i).$$

Clearly $w_i(x) > 0$ if $x \in U_i$ and $w_i(x) = 0$ if $x \notin U_i$. For each x at least one of the $w_i(x)$ is positive, since $X \subset \cup_j U_j$; and no more than $n+1$ are positive, since the covering $\{U_j\}$ is of order $\le n$. Set

$$\varphi_i(x) = \frac{w_i(x)}{\sum_{j=1}^r w_j(x)};$$

as with $w_i(\cdot)$, $\varphi_i(x) \ge 0$, since $\varphi_i(x) \ne 0$ iff $x \in U_i$, for each $x \in X$ at least if and only if one and no more than $n+1$ of $\{\varphi_j(x)\}$ are nonzero, and

$$\sum_{j=1}^r \varphi_j(x) = 1 \qquad \text{for every} \quad x \in X.$$

We now set[2]

$$g(x) = \sum_{i=1}^r \varphi_i(x)\, p_i. \tag{1.4}$$

Since the only nonzero terms in the sum are for values of i for which $x \in U_i$, it follows from (1.2) and (1.3) that for such values of i, $|p_i - f(x)| < \eta$ and hence

$$|g(x) - f(x)| = \left| \sum_{i=1}^r \varphi_i(x)(p_i - f(x)) \right| \le \sum_{i=1}^r \varphi_i(x)|p_i - f(x)| < \eta$$

for all $x \in X$.

[2] In fact g maps X into an n-dimensional polyhedron. Since no more than $n+1$ of the φ_ks are nonzero at any one time, for every $x \in U_i$ the image $g(x)$ is contained in some fixed simplex S_i of dimension $\le n$. Then $g(X) \subset \cup_{i=1}^r S_i$, where the right-hand side defines a polyhedron of dimension $\le n$. This remark will prove useful later in the proof of Theorem 2.12.

Now if $g(x) = g(x')$ then

$$g(x) - g(x') = \sum_{k=1}^{r} \underbrace{[\varphi_k(x) - \varphi_k(x')]}_{c_k} p_k = 0,$$

i.e.

$$\sum_{k=1}^{r} c_k p_k = 0 \qquad \text{with} \quad \sum_{k=1}^{r} c_k = \sum_{k=1}^{r} \varphi_k(x) - \sum_{k=1}^{r} \varphi_k(x') = 1 - 1 = 0.$$

We have a vanishing linear combination of the p_ks, with coefficients that sum to zero. Since no more than $n+1$ of the φ_ks are nonzero for each $x \in X$, no more than $2n+2$ of the c_ks are nonzero. Since the $\{p_k\}$ are in general position in $\mathbb{R}^{2n+1}$, any subcollection of the $\{p_k\}$ with $\leq 2n+2$ elements must be geometrically independent: it follows that $\varphi_k(x) = \varphi_k(x')$ for every $k = 1, \ldots, r$. In particular, since $x \in U_i$ for some i, $\varphi_i(x) > 0$; therefore $\varphi_i(x') > 0$, and hence $x' \in U_i$ too.

Thus $x, x' \in U_i$; since $\text{diam}(U_i) < \delta < \epsilon$ it follows that $g \in \mathscr{F}_\epsilon$. □

We can now use the Baire Category Theorem to show that a 'large' class of functions in $C(X, \mathbb{R}^{2n+1})$ are homeomorphisms. In common terminology, we show that in fact such functions are 'generic', meaning that they are a dense G_δ (a countable intersection of open sets).

Theorem 1.12 *Let X be a compact metric space with $\dim(X) \leq n$. Then there is a dense G_δ of functions in $C(X, \mathbb{R}^{2n+1})$ that are homeomorphisms of X into $\mathbb{R}^{2n+1}$.*

Proof For each $j \in \mathbb{N}$, $\mathscr{F}_{1/j}$ is open and dense in $C(X, \mathbb{R}^{2n+1})$. It follows from the Baire Category Theorem (Theorem 1.6) that $\cap_j \mathscr{F}_{1/j}$ is a dense G_δ in this space. But by Lemma 1.8 this is precisely the collection of all homeomorphisms of X into $\mathbb{R}^{2n+1}$. □

An example due to Flores (1935) shows that this result cannot be improved: the collection of all faces of a $2n+2$-dimensional cell (see Section V.9 in Hurewicz & Wallman (1941)) that have dimension $\leq n$ form an n-dimensional space which cannot be embedded into $\mathbb{R}^{2n}$ (see also Exercise 1.11.F in Engelking (1978)).

1.4 Large and small inductive dimensions

The small inductive dimension, $\text{ind}(\cdot)$, is defined as follows, where we use ∂U to denote the boundary of U:

(i) the empty set has $\text{ind}(\emptyset) = -1$;

(ii) $\operatorname{ind}(X) \leq n$ if for every point $p \in X$, p has arbitrarily small neighbourhoods U with $\operatorname{ind}(\partial U) \leq n-1$;
(iii) $\operatorname{ind}(X) = n$ if $\operatorname{ind}(X) \leq n$ but it is not true that $\operatorname{ind}(X) \leq n+1$.

The argument showing that the small inductive dimension and the covering dimension are equal in separable metric spaces is outlined in Exercises 1.5–1.7.

The large inductive dimension, $\operatorname{Ind}(\cdot)$, is defined similarly, but with (ii) replaced by

(ii′) $\operatorname{Ind}(X) \leq n$ if for every closed set $A \subset X$ and each open set $V \subset X$ that contains the set A there exists an open set $U \subset X$ such that

$$A \subset U \subset V \qquad \text{and} \qquad \operatorname{Ind}(\partial U) \leq n-1.$$

The large inductive dimension and the covering dimension coincide in any metric space (in fact, in any metrisable space), see Theorem 4.1.3 in Engelking (1978); clearly $\operatorname{ind}(X) \leq \operatorname{Ind}(X)$ always.

Exercises

1.1 Suppose that for every open cover $\{U_1, \ldots, U_{n+2}\}$ of X, there exists a cover of X by closed sets $\{F_1, \ldots, F_{n+2}\}$, with $F_j \subseteq U_j$ and $\cap_{j=1}^{n+2} F_j = \emptyset$. Show that $\dim(X) \leq n$. [Hint: first show that the assumption implies that the same is true with the $\{F_j\}$ open.] (Theorem 3.2.1 in Edgar (2008) shows that in fact the assumption here and $\dim(X) \leq n$ are equivalent.)

1.2 Find a covering of $\mathbb{R}^2$ of order 3 and mesh size no larger than 1. Deduce that any compact subset X of $\mathbb{R}^2$ has $\dim(X) \leq 2$. [Hint: any open covering of X has a Lebesgue number that is strictly positive, see the proof of Lemma 1.7.]

1.3 Let A_1, A_2, and B be mutually disjoint subsets of a space X. We say that B separates A_1 and A_2 in X if there exist two disjoint sets U_1 and U_2, open in X, such that

$$A_1 \subset U_1, \quad A_2 \subset U_2, \quad \text{and} \quad X \setminus B \subset U_1 \cup U_2.$$

Now let A be a closed subset of X, C and C' a pair of disjoint closed subsets of X, and K a closed subset of A that separates $A \cap C$ and $A \cap C'$ in A. Show that there exists a closed set B that separates C and C' in X and satisfies $A \cap B \subset K$.

1.4 Show that a one-to-one continuous mapping of a compact set is a homeomorphism.

1.5 Assume that if M is a subspace of (X, ϱ) with $\mathrm{ind}(M) \leq 0$ then given any two open sets U_1 and U_2 that cover M, there exist disjoint open sets V_1 and V_2 with $V_1 \subset U_1$ and $V_2 \subset U_2$ such that V_1 and V_2 still cover M. Use induction to show that if $\{U_1, \ldots, U_r\}$ is an open cover of M then there exists an open cover $\{V_1, \ldots, V_r\}$ of M such that

$$V_j \subset U_j \qquad \text{and} \qquad V_i \cap V_j = \emptyset \quad \text{for} \quad i \neq j;$$

i.e. that $\mathrm{ind}(M) \leq 0$ implies that $\dim(M) = 0$.

1.6 As mentioned immediately before the statement of Proposition 1.3, a fundamental result in the theory of the small inductive dimension is that a set $A \subseteq (X, \varrho)$ has $\mathrm{ind}(A) \leq n$ if and only if it is the union of $n + 1$ subspaces of dimension ≤ 0. Use this result along with that of the previous exercise to show that $\dim(A) \leq \mathrm{ind}(A)$.

1.7 Deduce from the following three facts that $\dim(X) = \mathrm{ind}(X)$ for any separable metric space (reference is given to the relevant results in Hurewicz & Wallman (1941)):

(i) any separable metric space X with $\dim(X) \leq n$ can be embedded into $\mathcal{M}^n_{2n+1} \cap I_{2n+1}$, the set of points in I_{2n+1} at most n of whose coordinates are rational (Theorem V.5);

(ii) $\mathrm{ind}(\mathcal{M}_{2n+1}) = n$ (Example IV.1); and

(iii) $A \subseteq B$ implies that $\mathrm{ind}(A) \leq \mathrm{ind}(B)$ (Theorem III.1).

2

Hausdorff measure and Hausdorff dimension

The Hausdorff dimension, which we denote by d_{H}, is one of the most widely used definitions. It finds extensive application in geometric measure theory (see Federer (1969), for example), and in the theory of dynamical systems (see Pesin (1997), Boichenko, Leonov, & Reitmann (2005)). Much of its power is due to the fact that it is defined in terms of Hausdorff measures, naturally linking dimension and measure.

It occupies an intermediate position between the covering dimension and the box-counting dimension (d_{B}), with $\dim(X) \le d_{\mathrm{H}}(X) \le d_{\mathrm{B}}(X)$ (Theorem 2.11 and Lemma 3.3(v)). Since $\dim(X) \le d_{\mathrm{H}}(X)$ we can use Theorem 1.12 to guarantee that any set with finite Hausdorff dimension can be topologically embedded into a Euclidean space. However, we will see at the beginning of Chapter 6 that there are examples of sets with finite Hausdorff dimension that cannot be embedded into a Euclidean space using any map that is linear.

2.1 Hausdorff measure and Lebesgue measure

Although we will ultimately consider subsets of Banach spaces, we begin in a relatively abstract way by defining the s-dimensional Hausdorff measure $\mathscr{H}^s$, and the Hausdorff dimension d_{H}, for subsets of a metric space (X, ϱ).

An outer measure μ on X assigns a nonnegative real number to every subset of X, with the properties

(i) $\mu(\emptyset) = 0$;
(ii) if $A \subseteq B$ then $\mu(A) \le \mu(B)$; and
(iii) if $\{A_j\}_{j=1}^{\infty}$ are subsets of X then

$$\mu\left(\bigcup_{j=1}^{\infty} A_j\right) \le \sum_{j=1}^{\infty} \mu(A_j). \tag{2.1}$$

We do not distinguish in what follows between 'outer measures' and 'measures'; strictly speaking a 'measure' is only defined on a σ-algebra of measurable sets (sets for which $\mu(E) = \mu(E \cap A) + \mu(E \setminus A)$), but any measure can be extended to an outer measure, and any outer measure gives rise to a measure when restricted to the σ-algebra of measurable sets (for more details see Chapter 1 of Mattila (1995), or of Rogers (1998)). A probability measure on (X, ϱ) is a measure μ with $\mu(X) = 1$.

We now define an approximation to the s-dimensional Hausdorff measure, and show that it is an outer measure. For a subset U of X, we recall that $|U| = \sup_{x,y\in U} \varrho(x, y)$; for $A \subseteq X$, $s \geq 0$, and $\delta > 0$, we define

$$\mathscr{H}_\delta^s(A) = \inf \left\{ \sum_{i=1}^{\infty} |U_i|^s \ : \ X \subseteq \cup_{i=1}^{\infty} U_i \text{ with } |U_i| \leq \delta \right\}.$$

Note that any sets $\{U_i\}$ are allowable in this cover of X (they need not be open).

Lemma 2.1 *$\mathscr{H}_\delta^s$ in an outer measure on (X, ϱ) for each $\delta > 0$.*

Proof Fix $\delta > 0$. Clearly (i) $\mathscr{H}_\delta^s(\emptyset) = 0$ and (ii) $\mathscr{H}_\delta^{(s)}(A) \leq \mathscr{H}_\delta^s(B)$ whenever $A \subseteq B$. To prove (iii) let $\{A_j\}_{j=1}^{\infty}$ be a collection of subsets of X. Given $\epsilon > 0$ there exists a sequence $\{B_j^{(i)}\}_{i=1}^{\infty}$ of subsets of X such that

$$A_j \subset \bigcup_{i=1}^{\infty} B_j^{(i)}, \quad |B_j^{(i)}| \leq \delta, \quad \text{and} \quad \sum_{i=1}^{\infty} |B_j^{(i)}|^s \leq \mathscr{H}_\delta^s(A_j) + \epsilon 2^{-j}.$$

It follows that

$$\bigcup_{j=1}^{\infty} A_j \subset \bigcup_{i,j=1}^{\infty} B_j^{(i)} \quad \text{and} \quad \mathscr{H}_\delta^s \left(\bigcup_{j=1}^{\infty} A_j \right) \leq \sum_{i,j=1}^{\infty} |B_j^{(i)}|^s \leq \epsilon + \sum_{j=1}^{\infty} \mathscr{H}_\delta^s(A_j).$$

Since this is valid for any $\epsilon > 0$, $\mathscr{H}_\delta^s\left(\bigcup_{j=1}^{\infty} A_j\right) \leq \sum_{j=1}^{\infty} \mathscr{H}_\delta^s(A_j)$ as required. □

One obtains the s-dimensional Hausdorff measure by refining the cover involved in the definition of $\mathscr{H}_\delta^s$, i.e. taking the limit as $\delta \to 0$:

$$\mathscr{H}^s(X) = \lim_{\delta \to 0} \mathscr{H}_\delta^s(X).$$

The limit exists (it may be infinity) since $\mathscr{H}_\delta^s(X)$ increases as δ decreases. It follows immediately from Lemma 2.1 that $\mathscr{H}^s$ is an outer measure; in fact more is true.

Theorem 2.2 *$\mathscr{H}^s$ is a metric outer measure, i.e.*

$$\mathscr{H}^s(A \cup B) = \mathscr{H}^s(A) + \mathscr{H}^s(B) \tag{2.2}$$

whenever $\operatorname{dist}(A, B) > 0$.

Proof We have already remarked that $\mathcal{H}^s$ is an outer measure, so we have only to prove (2.2). Let A, B be subsets of X such that $\operatorname{dist}(A, B) > 0$ and let $\delta < \operatorname{dist}(A, B)$. It is easy to see that

$$\mathcal{H}^s_\delta(A \cup B) = \mathcal{H}^s_\delta(A) + \mathcal{H}^s_\delta(B).$$

Taking $\delta \to 0$ yields $\mathcal{H}^s(A \cup B) = \mathcal{H}^s(A) + \mathcal{H}^s(B)$, as required. □

A set $A \subset X$ is said to be $\mathcal{H}^s$-measurable if for each $E \subset X$

$$\mathcal{H}^s(E) = \mathcal{H}^s(E \cap A) + \mathcal{H}^s(E \cap A^c).$$

Since $\mathcal{H}^s$ is a metric outer measure, it follows (for a proof see Theorem 1.5 in Falconer (1985)) that every closed subset of X is $\mathcal{H}^s$-measurable; and hence that every Borel subset of X is $\mathcal{H}^s$-measurable. In particular, if $\{A_j\}$ are disjoint Borel sets then

$$\mathcal{H}^s\left(\bigcup_{j=1}^{\infty} A_j\right) = \sum_{j=1}^{\infty} \mathcal{H}^s(A_j),$$

i.e. equality holds in (2.1).

We end this section with the result that for subsets of $\mathbb{R}^n$, $\mathcal{H}^n$ is a constant multiple of n-dimensional Lebesgue measure. We will require the fact that the volume of any subset of $\mathbb{R}^n$ is no larger than the volume of a ball with the same diameter; for a proof see Section 2.2 in Evans & Gariepy (1992). We use $\mathcal{L}^n$ to denote n-dimensional Lebesgue measure.

Theorem 2.3 *For any bounded subset A of $\mathbb{R}^n$,*

$$\mathcal{L}^n(A) \le \Omega_n \left(\frac{|A|}{2}\right)^n,$$

where $\Omega_n = \pi^{n/2}/\Gamma(n/2 + 1)$ is the volume of the unit ball in $\mathbb{R}^n$.

Theorem 2.4 *If A is a bounded subset of $\mathbb{R}^n$ then*

$$\mathcal{H}^n(A) = 2^{-n}\Omega_n \mathcal{L}^n(A).$$

Proof First, given any $\epsilon > 0$ cover A by sets $\{U_i\}$ such that

$$\sum_i |U_i|^n < \mathcal{H}^n(A) + \epsilon.$$

Using Theorem 2.3,

$$\mathcal{L}^n(A) \le \sum_i \mathcal{L}^n(U_i) \le \sum_i \Omega_n \left(\frac{|U_i|}{2}\right)^n < 2^{-n}\,\Omega_n[\mathcal{H}^n(A) + \epsilon],$$

which implies that $\mathscr{L}^n(A) \leq 2^{-n}\Omega_n \mathscr{H}^n(A)$. The lower bound relies on the Vitali Covering Theorem, and since we only make use of the upper bound in what follows we refer to Falconer (1985) for both the covering theorem (his Theorem 1.10) and the proof of the lower bound (his Theorem 1.12). □

2.2 Hausdorff dimension

We now show that there is a 'critical value' of s at which the s-dimensional Hausdorff measure switches (as s is decreased) from being zero to being infinite – this will be how we define the Hausdorff dimension.

Proposition 2.5 *Let A be a subset of (X, ϱ). Take $s' > s > 0$: if $\mathscr{H}^s(A) < \infty$ then $\mathscr{H}^{s'}(A) = 0$, and if $\mathscr{H}^{s'}(A) > 0$ then $\mathscr{H}^s(A) = \infty$.*

Proof The two statements are equivalent; we prove the first. If $\mathscr{H}^s(A) < \infty$ then for any $\delta > 0$ there is a cover of A by sets $\{B_j\}$ with diameters $\leq \delta$ such that

$$\sum_{j=1}^{\infty} |B_j|^s \leq \mathscr{H}^s(A) + 1.$$

It follows that for $s' > s$

$$\sum_{j=1}^{\infty} |B_j|^{s'} \leq \delta^{s'-s} \sum_{j=1}^{\infty} |B_j|^s \leq \delta^{s'-s}[\mathscr{H}^s(A) + 1],$$

and hence $\mathscr{H}^{s'}(A) = 0$. □

We can now define the Hausdorff dimension.

Definition 2.6 For any $A \subseteq (X, \varrho)$, the Hausdorff dimension of A is

$$d_{\mathrm{H}}(A) = \inf\{d \geq 0 : \mathscr{H}^d(A) = 0\}.$$

In the light of this definition, the following simple lemma will be useful.

Lemma 2.7 *If $A \subseteq X$ and $s > 0$ then $\mathscr{H}^s(A) = 0$ if and only if for every $\epsilon > 0$ there is a countable covering of A, $\{U_j\}_{j=1}^{\infty}$, such that*

$$\sum_{j=1}^{\infty} |U_j|^s < \epsilon. \tag{2.3}$$

Proof Suppose that $\mathscr{H}^s(A) = 0$. Then for any $\delta > 0$ there is a cover of A by sets $\{U_j\}$ with $|U_j| < \delta$ such that (2.3) holds; in particular one such cover exists. Conversely, given any $\delta > 0$, choose $\epsilon > 0$ such that $\epsilon^{1/s} < \delta$, and find a covering that satisfies (2.3). Then this must be a covering by sets with $|U_j| < \delta$ that satisfies (2.3), and hence $\mathscr{H}^s(A) < \epsilon$ for every $\epsilon > 0$, i.e. $\mathscr{H}^s(A) = 0$. □

We now prove some basic properties of the Hausdorff dimension.

Proposition 2.8

(i) *If $A, B \subseteq (X, \varrho)$ and $A \subseteq B$ then $d_{\rm H}(A) \le d_{\rm H}(B)$;*

(ii) *the Hausdorff dimension is stable under countable unions: if $X_k \subseteq X$ then*

$$d_{\rm H}\left(\bigcup_{k=1}^{\infty} X_k\right) = \sup_k d_{\rm H}(X_k); \tag{2.4}$$

(iii) *if U is an open subset of $\mathbb{R}^n$ then $d_{\rm H}(U) = n$, in particular $d_{\rm H}(\mathbb{R}^n) = n$; and*

(iv) *if $f : (X, \varrho_X) \to (Y, \varrho_Y)$ is Hölder continuous with exponent $\theta \in (0, 1]$,*

$$\varrho_Y(f(x_1), f(x_2)) \le C\, \varrho_X(x_1, x_2)^{\theta},$$

then $d_{\rm H}(f(X)) \le d_{\rm H}(X)/\theta$.

Proof (i) The proof is immediate from the definition.

(ii) If $\sup_k d_{\rm H}(X_k) = \infty$ then it follows from (i) that $d_{\rm H}(X) = \infty$. So we can assume that $\sup_k d_{\rm H}(X_k) < \infty$, and take $s > \sup_k d_{\rm H}(X_k)$: then $\mathscr{H}^s(X_k) = 0$ for every k, and since $\mathscr{H}^s$ is an outer measure (2.1) $\mathscr{H}^s(\cup_k X_k) = 0$ and hence $d_{\rm H}(X) < s$, from which (2.4) follows.

(iii) By considering $U = \cup_{j=1}^{\infty}[U \cap B(0, j)]$ and using (ii), it suffices to show that $d_{\rm H}(U) = n$ for any bounded open set U. Certainly U is contained in some cube C with sides of length R. Given $\delta > 0$, choose $k \in \mathbb{N}$ such that $k > R\sqrt{n}/\delta$, and divide C into k^n subcubes with sides of length R/k (and so with diameters $\sqrt{n}R/k < \delta$); then $\mathscr{H}^n_\delta(C) \le R^n k^n (k^{-1}\sqrt{n})^n \le R^n n^{n/2}$ and it follows that $\mathscr{H}^n(C) < \infty$, whence $d_{\rm H}(C) \le n$. It follows from part (i) that $d_{\rm H}(U) \le n$. To show the lower bound, it follows from Theorem 2.4 that $\mathscr{H}^n(U) \ge 2^n \Omega_n^{-1} \mathscr{L}^n(U) > 0$, and hence $d_{\rm H}(U) \ge n$.

(iv) Take $s > d_{\rm H}(X)$. Then for any $\epsilon > 0$ there exists a cover $\{U_j\}$ of X with

$$\sum_j |U_j|^s < \epsilon.$$

Then $\{f(U_j)\}$ is a cover of $f(X)$ and $|f(U_j)| \le C|U_j|^{\theta}$, from which it follows that

$$\sum_j |f(U_j)|^{s/\theta} < C^s \epsilon.$$

Lemma 2.7 guarantees that $\mathscr{H}^{s/\theta}(f(X)) = 0$, and hence $d_{\mathrm{H}}(f(X)) \leq s/\theta$. □

Note that it is immediate from (ii) that the Hausdorff dimension of any countable set is zero. In subsequent chapters we will frequently have recourse to the example of an orthogonal sequence in a Hilbert space,

$$\{a_j e_j\}_{j=1}^{\infty} \cup \{0\}$$

where $|a_j| \to 0$ and $\{e_j\}_{j=1}^{\infty}$ is an orthonormal set. In the light of this remark, any such set will have zero Hausdorff dimension.

2.3 The Hausdorff dimension of products

It is generally hard to find lower bounds on the Hausdorff dimension, but the following powerful theorem ('Frostman's Lemma') is very useful for this. We will only prove the implication in one direction, which is easy. The argument to prove the converse is very involved, see Mattila (1995, Theorem 8.8; he also gives a proof, due to Howroyd (1995), valid in compact metric spaces).

Theorem 2.9 *Let X be a closed subset of $\mathbb{R}^n$. Then $\mathscr{H}^s(X) > 0$ if and only if there exists a probability measure μ supported on X such that*

$$\mu(B(x,r)) \leq cr^s \qquad \text{for all} \qquad x \in X,\ r > 0.$$

Proof We only prove the 'if' part. Take any cover $\{B_{r_i}(x_i)\}$ of X with $r_i \leq \delta$; then

$$1 = \mu(X) = \mu\left(\bigcup_i X \cap B_{r_i}(x_i)\right) \leq \sum \mu\left(X \cap B_{r_i}(x_i)\right) \leq c \sum_i r_i^s.$$

Taking the infimum, it follows that $\mathscr{H}^s(X) \geq 1/c$. □

Using this result we can show that $d_{\mathrm{H}}(X \times Y) \geq d_{\mathrm{H}}(X) + d_{\mathrm{H}}(Y)$. While the result remains true in greater generality, since we use Theorem 2.9 we state it for subsets of Euclidean spaces. We take the norm on $\mathbb{R}^n \times \mathbb{R}^m$ to be the standard norm on $\mathbb{R}^{n+m}$.

Proposition 2.10 *Let $X \subset \mathbb{R}^n$ and $Y \subset \mathbb{R}^m$ be closed sets. Then*

$$d_{\mathrm{H}}(X \times Y) \geq d_{\mathrm{H}}(X) + d_{\mathrm{H}}(Y).$$

Proof Given $s < d_{\mathrm{H}}(X)$ and $t < d_{\mathrm{H}}(Y)$, $\mathscr{H}^s(X) > 0$ and $\mathscr{H}^t(Y) > 0$. It follows from Theorem 2.9 that there exist probability measures μ and ν supported on X and Y respectively such that

$$\mu(B(x,r)\cap X)\leq c_1 r^s \qquad \text{and} \qquad \nu(B(y,r)\cap Y)\leq c_2 r^t.$$

Since $B((x,y),r)\subset B(x,r)\times B(y,r)$, the measure $\mu\times\nu$ on $\mathbb{R}^n\times\mathbb{R}^m$ satisfies $(\mu\times\nu)(X\times Y)=\mu(X)\nu(Y)=1$ and

$$(\mu\times\nu)(B((x,y),r)\cap(X\times Y))\leq\mu(B(x,r)\cap X)\nu(B(y,r)\cap Y)\leq c_1c_2 r^{s+t}.$$

It follows that $\mathscr{H}^{s+t}(X\times Y)>0$, and hence $d_{\rm H}(X\times Y)\geq s$. □

The reverse inequality does not hold in general, as the following example shows (Theorem 5.11 in Falconer (1985)). Let $m_0=1$, $m_{k+1}=k\sum_{j=0}^k m_j$.

Let X consist of those numbers in $[0,1]$ that have a zero in the rth decimal place for $m_k+1\leq r\leq m_{k+1}$ and k even, and let Y consist of the numbers in $[0,1]$ with a zero in the rth decimal place for $m_k+1\leq r\leq m_{k+1}$ and k odd.

Each of these sets X and Y has Hausdorff dimension zero. For X, consider the first m_{k+1} decimal places, with k even; X can be covered by 10^{n_k} intervals $\{I_j\}$ of length $10^{-m_{k+1}}$, where

$$n_k=(m_2-m_1)+(m_4-m_3)+\cdots+(m_k-m_{k-1}).$$

Then

$$\sum_j |I_j|^{1/k}\leq 10^{n_k}\times 10^{-m_{k+1}/k}\leq 10^{\sum_{j=0}^k m_j}\times 10^{-\sum_{j=0}^k m_j}=1,$$

by the choice of m_k. It follows that $d_{\rm H}(X)=0$. A similar argument shows that $d_{\rm H}(Y)=0$.

However, any $z\in(0,1)$ can be written in the form $z=x+y$ with $x\in X$ and $y\in Y$, and the mapping $f:X\times Y\to\mathbb{R}$ given by $(x,y)\mapsto x+y$ is Lipschitz. It follows that

$$1=d_{\rm H}(0,1)\leq d_{\rm H}(f(X\times Y))\leq d_{\rm H}(X\times Y)$$

using Proposition 2.8(iv).

2.4 Hausdorff dimension and covering dimension

We are now in a position to show that the Hausdorff dimension bounds the covering dimension. The proof given here, which works directly with the covering dimension, rather than the small inductive dimension as in Theorem VII 3 of

Hurewicz & Wallman (1941), is due to Edgar (2008, Theorem 6.3.11). A similar argument, valid in any separable metric space, is given by Charalambous (1999).

Note that an immediate consequence of this result is that $\dim(X) \le n$ for any compact subset of $\mathbb{R}^n$, since Proposition 2.8(iii) shows that $d_H(\mathbb{R}^n) = n$, and d_H is monotonic (part (i) of the same proposition), so that $d_H(X) \le n$ for any subset of $\mathbb{R}^n$.

Theorem 2.11 *Let X be a compact metric space. Then* $\dim(X) \le d_H(X)$.

Proof We use the characterisation of covering dimension from Exercise 1.1. Let $n = \dim(X)$, so that it is not true that $\dim(X) \le n - 1$. Then there must exist an open cover $\{U_i\}_{i=1}^{n+1}$ of X such that for any closed sets $\{F_i\}$ with $F_i \subset U_i$ that still form a cover of X, $\cap_{i=1}^{n+1} F_i \neq \emptyset$.

Now define

$$\delta_i(x) = \operatorname{dist}(x, X \setminus U_i) \qquad i = 1, \ldots, n+1$$

and $\delta(x) = \delta_1(x) + \cdots + \delta_{n+1}(x)$. Then each δ_i is Lipschitz continuous, and hence so is δ:

$$|\delta_i(x) - \delta_i(y)| \le \varrho(x, y) \qquad \text{and} \qquad |\delta(x) - \delta(y)| \le (n+1)\varrho(x, y).$$

Since the $\{U_i\}$ form a cover of X, $x \in U_i$ for some i, and so $\delta_i(x) > 0$; it follows that $\delta(x) > 0$ for every $x \in X$, and so since X is compact, there exist $b > a > 0$ such that $a \le \delta(x) \le b$ for every $x \in X$. Define $h : X \to \mathbb{R}^{n+1}$ by

$$h(x) = \left(\frac{\delta_1(x)}{\delta(x)}, \frac{\delta_2(x)}{\delta(x)}, \ldots, \frac{\delta_{n+1}(x)}{\delta(x)} \right).$$

The function h is again Lipschitz, since

$$\begin{aligned} \left| \frac{\delta_j(x)}{\delta(x)} - \frac{\delta_j(y)}{\delta(y)} \right| &= \frac{|\delta(y)\delta_j(x) - \delta(x)\delta_j(y)|}{\delta(x)\delta(y)} \\ &\le a^{-2}[\delta(y)|\delta_j(x) - \delta_j(y)| + \delta_j(y)|\delta(y) - \delta(x)|] \\ &\le a^{-2}b(n+2)\,\varrho(x, y), \end{aligned}$$

and so

$$|h(x) - h(y)| \le a^{-2}b(n+2)\sqrt{n}\,\varrho(x, y).$$

Now, since h is Lipschitz, $d_H(h(X)) \le d_H(X)$ (Proposition 2.8(iv)). The proof is concluded by showing that $h(X)$ contains the simplex

$$T = \{(t_1, \ldots, t_{n+1}) \in \mathbb{R}^{n+1} : \ t_i > 0 \text{ and } \sum_{i=1}^{n+1} t_i = 1\},$$

which clearly has $d_{\rm H}(T) \geq n$ since it is bi-Lipschitz equivalent to an open subset of $\mathbb{R}^n$. To this end, take a point $\underline{t} = (t_1, \ldots, t_{n+1}) \in T$ and consider the sets

$$F_i = \left\{ x \in X : \frac{\delta_i(x)}{\delta(x)} \geq t_i \right\}.$$

Each F_i is closed, $F_i \subseteq U_i$, and the $\{F_i\}$ form a cover of X since

$$\sum_{i=1}^{n+1} [\delta_i(x)/\delta(x)] = 1.$$

Since $\cap_{i=1}^{n+1} F_i \neq \emptyset$, there exists an $x \in X$ with $\delta_i(x)/\delta(x) \geq t_i$ for each i. But $\sum_i [\delta_i(x)/\delta(x)] = 1$ and $\sum_i t_i = 1$, whence it follows that $\delta_i(x)/\delta(x) = t_i$, i.e. that $h(x) = \underline{t}$, and so $h(X) \supseteq T$. □

Of course, this inequality can be strict, since $\dim(X)$ is an integer-valued definition of dimension, and there exist sets for which $d_{\rm H}(X) \notin \mathbb{N}$. However, we always have equality for some homeomorphic image of X:

Theorem 2.12 *If (X, ϱ) is compact and* $\dim(X) = n$ *then there is a homeomorphism $h : X \to \mathbb{R}^{2n+1}$ such that $d_{\rm H}(h(X)) = n$.*

The proof is taken from Hurewicz & Wallman (1941, Theorem VII.4).

Proof Take $s > n$, and consider the collection K_s of all those functions $f \in C(X, \mathbb{R}^{2n+1})$ for which $\mathscr{H}^s(f(X)) = 0$. The condition that $\mathscr{H}^s(f(X)) = 0$ means that for each $i \in \mathbb{N}$ there exists a finite cover $\{X_j\}_{j=1}^k$ of X such that

$$\sum_{j=1}^{k} |f(X_j)|^s < 1/i. \tag{2.5}$$

Let $\mathbb{X}$ denote a finite cover $\{X_j\}_{j=1}^k$ of X, and denote by $G_{i,s}^{\mathbb{X}}$ the set of all functions $f \in C(X, \mathbb{R}^{2n+1})$ that satisfy (2.5) for this decomposition. Then

$$K_s = \bigcap_{i=1}^{\infty} \left[\bigcup_{\text{all possible } \mathbb{X}} G_{i,s}^{\mathbb{X}} \right].$$

Since each $G_{i,s}^{\mathbb{X}}$ is an open subset of $C(X, \mathbb{R}^{2n+1})$, so is the expression in square brackets for each i. It follows that K_s is a G_δ in $C(X, \mathbb{R}^{2n+1})$.

Now, as noted during the proof of the embedding theorem for sets with $\dim(X)$ finite, the embedding map g defined in (1.4) maps X into an n-dimensional polyhedron. Thus the set of maps $f \in C(X, \mathbb{R}^{2n+1})$ that map X

into such a polyhedron is dense; for any such map,

$$d_{\mathrm{H}}(f(X)) \leq d_{\mathrm{H}}(\text{polyhedron}) \leq n.$$

Since all such maps must therefore lie in K_s, K_s contains a dense subset of $C(X, \mathbb{R}^{2n+1})$, so is itself dense.

We have shown that K_s is a dense G_δ in $C(X, \mathbb{R}^{2n+1})$, and Theorem 1.12 guarantees that the set of maps E_X that are embeddings of X is also a dense G_δ. It follows from the Baire Category Theorem (Theorem 1.6) that

$$E_X \cap \bigcap_{j\geq 1} K_{n+(1/j)}$$

is also a dense G_δ. In particular this set is nonempty, so there exists a homeomorphism f of X into $\mathbb{R}^{2n+1}$ such that $\mathscr{H}^s(X) = 0$ for all $s \geq n$, i.e. for which $d_{\mathrm{H}}(f(X)) \leq n$. □

As a corollary, we give what amounts to an alternative definition of the covering dimension. It is immediate from this definition that dim is a topological invariant, but in no way clear that dim must be an integer (cf. Prosser, 1970).

Corollary 2.13 *If X is a compact space then*

$$\dim(X) = \inf\{d_{\mathrm{H}}(X') : X' \text{ homeomorphic to } X\}.$$

Mandelbrot (1982) defined a 'fractal' as a set for which $\dim(X) < d_{\mathrm{H}}(X)$. Luukkainen (1998) makes the nice comment that the result of this corollary implies that 'there is no purely topological reason for X to be fractal'.

Exercises

2.1 Let X be a subset of $\mathbb{R}^n$, and let $f : X \to \mathbb{R}^m$ satisfy

$$|f(x_1) - f(x_2)| \leq C|x_1 - x_2|^\theta,$$

with $C > 0$ and $\theta \in [0, 1]$. Show that the Hausdorff dimension of the graph

$$G = \{(x, f(x)) : x \in X\} \subset \mathbb{R}^{n+m}$$

is less than or equal to $n + (1 - \theta)m$.

2.2 Show that if J_q are open subsets of $[0, 1]$ such that

$$\sum |J_q| = 1 \qquad \text{and} \qquad \sum_q |J_q|^s < \infty, \tag{2.6}$$

then $[0, 1] \setminus \cup J_q$ has s-dimensional Hausdorff measure zero. (This result has applications to bounding the set of singular times in weak solutions of the three-dimensional Navier–Stokes equations, see Scheffer (1976).)

2.3 Define the 'd-dimensional spherical Hausdorff measure' of a set X by $\mathscr{S}^d(X) = \lim_{\delta\to 0} \mathscr{S}^d_\delta(X)$, where

$$\mathscr{S}^d_\delta(X) = \inf\left\{\sum r_i^d \,:\, X \subseteq \bigcup_{i=1}^{\infty} B(x_i, r_i) \,:\, r_i \le \delta\right\}.$$

Show that $d_{\rm H}(X) = \inf\{d \,:\, \mathscr{S}^d(X) = 0\}$.

2.4 Suppose that X is a bounded subset of $\mathbb{R}^n$ that is covered by a collection of balls $\{B(x, r(x))\}_{x\in X}$. Show that there exists a finite or countably infinite disjoint subcollection of this cover, $\{B(x_j, r(x_j))\}_{j=1}^{\infty}$, such that $\{B(x_j, 5r(x_j))\}_{j=1}^{\infty}$ still covers X. [Hint: set $M = \sup_{x\in X} r(x)$. Define

$$X_k = \{x \in X \,:\, (\tfrac{3}{4})^k M < r(x) \le (\tfrac{3}{4})^{k-1} M\},$$

and given $\{x_1, \ldots, x_{n_k}\}$ such that

$$\cup_{x\in X_k} B(x, r(x)) \subseteq \bigcup_{i=1}^{k_n} B(x_i, 5r(x_i)), \tag{2.7}$$

find points $\{x_{n_k+1}, \ldots, x_{n_{k+1}}\}$ in

$$X'_k = \{x \in X_k \,:\, B(x, r(x)) \cap \bigcup_{i=1}^{n_{k-1}} B(x_i, r(x_i)) = \emptyset\},$$

such that (2.7) holds with k replaced by $k + 1$.]

2.5 Suppose that $f \in L^1_{\rm loc}(\mathbb{R}^n)$. For some $0 \le d < n$ and some $\delta > 0$ define

$$S = \left\{x \in \mathbb{R}^n \,:\, \limsup_{r\to 0} \frac{1}{r^d} \int_{B(x,r)} |f(x)|\, {\rm d}x > \delta\right\}.$$

Use the results of the previous two exercises to show that $\mathscr{H}^d(S) = 0$. [Hint: first show that $\mathscr{L}^n(S) = 0$.] (This result forms the final piece in the proof of the partial regularity of the three-dimensional Navier–Stokes equations due to Caffarelli, Kohn, & Nirenberg (1982).)

3
Box-counting dimension

The study of the box-counting dimension forms the core of Part I of this book. We concentrate on the upper box-counting dimension (Definition 3.1), since this is the least restrictive definition of dimension that allows one to obtain a parametrisation of a 'finite-dimensional' set, using a finite number of parameters, that has a well-defined degree of continuity (Hölder). But it is also of interest since the upper box-counting dimension of many attractors arising in the infinite-dimensional dynamical systems is finite. We explore the implications of this fact in Part II.

As with the topological and Hausdorff dimensions, we give general results for subsets of metric spaces; but as we switch to particular examples (and then later embedding results) we specialise to subsets of Hilbert and Banach spaces.

3.1 The definition of the box-counting dimension

Let $N(X, \epsilon)$ denote the minimum number of balls of radius ϵ ('ϵ-balls') with centres in X required to cover X. We define the box-counting dimension of X as

$$d_{\text{box}}(X) = \lim_{\epsilon \to 0} \frac{\log N(X, \epsilon)}{-\log \epsilon}; \tag{3.1}$$

essentially $N(X, \epsilon) \sim \epsilon^{-d_{\text{box}}(X)}$ as $\epsilon \to 0$.

However, the limit in (3.1) need not exist in general, as the following example shows (cf. Exercise 3.8 in Falconer (1990)). Form a Cantor-like set $C = \cap_{j=1}^{\infty} C_j$, where C_j is the set at the end of stage j of the following construction: at stage $2j - 1$ remove the middle half (i.e. the two middle quarters) 2^{j-1} times, and at stage $2j$ remove the middle third 2^{j-1} times. By considering

C_{2j-1}, one can see that C requires

$$N_{2j-1} := 2^{2^j+2^{j-1}-2} \quad \text{intervals of length} \quad \epsilon_{2j-1} := 4^{-(2^j-1)}3^{-(2^{j-1}-1)},$$

to cover it; by considering C_{2j}, C requires

$$N_{2j} := 2^{2^{j+1}-2} \quad \text{intervals of length} \quad \epsilon_{2j} := 4^{-(2^j-1)}3^{-(2^j-1)}$$

for a cover. Thus

$$\frac{\log N_{2j-1}}{-\log \epsilon_{2j-1}} = \frac{(2^j + 2^{j-1} - 2)\log 2}{(2^j - 1)\log 4 + (2^{j-1} - 1)\log 3} \to \frac{3\log 2}{2\log 4 + \log 3},$$

while

$$\frac{\log N_{2j}}{-\log \epsilon_{2j}} = \frac{(2^{j+1} - 2)\log 2}{(2^j - 1)\log 4 + (2^j - 1)\log 3} \to \frac{2\log 2}{\log 4 + \log 3}.$$

We therefore make the following two definitions.

Definition 3.1 Let (X, ϱ) be a metric space, and $A \subseteq X$. Let $N(A, \epsilon)$ denote the minimum number of closed balls of radius ϵ with centres in A required to cover A. The upper box-counting dimension of A is

$$d_{\mathrm{B}}(A) = \limsup_{\epsilon \to 0} \frac{\log N(A, \epsilon)}{-\log \epsilon}, \tag{3.2}$$

and the lower box-counting dimension of A is

$$d_{\mathrm{LB}}(A) = \liminf_{\epsilon \to 0} \frac{\log N(A, \epsilon)}{-\log \epsilon}.$$

The inequality $d_{\mathrm{LB}}(A) \le d_{\mathrm{B}}(A)$ is clear, with the above Cantor-like set showing that it can be strict. The 'box-counting dimension' (3.1) only exists when the lower and upper box-counting dimensions coincide; but since we will (almost) always be interested in the upper box-counting dimension in what follows, we will usually refer to the quantity $d_{\mathrm{B}}(X)$ defined in (3.2) as the box-counting dimension.[1] (We will see in Section 8.2 that sets with finite lower box-counting dimension do not enjoy the same embedding properties as sets with finite upper box-counting dimension.)

It is immediate from the definition that if $d > d_{\mathrm{B}}(A)$ then there exists an ϵ_0 such that

$$N(A, \epsilon) < \epsilon^{-d} \qquad \text{for all} \qquad \epsilon < \epsilon_0, \tag{3.3}$$

[1] In much of the dynamical systems literature the upper box-counting dimension is referred to as the 'fractal dimension'; although a little inelegant, 'box-counting dimension' is to be preferred for obvious reasons.

while if $d < d_{\rm B}(A)$ then there exists a sequence $\epsilon_j \to 0$ such that

$$N(A, \epsilon) > \epsilon_j^{-d}.$$

We will often make use of these (particularly (3.3)) in what follows.

Note that we could just as well take a covering by open balls, since any covering by open balls of radius ϵ yields a covering by closed balls of the same radius, and any covering by closed balls of radius ϵ yields a covering by open balls of radius 2ϵ. A number of other alternative, but equivalent, definitions are discussed in Exercise 3.1.

Sometimes it is useful to be able to calculate the box-counting dimension by taking the limit (superior) through a sequence $\{\varepsilon_k\}$ of values of ϵ, rather than a continuous limit. The Cantor set example above shows that one cannot do this without imposing some restrictions on ε_k, such as those in the following lemma, whose main application is to the geometric sequence $\varepsilon_k = c\alpha^k$.

Lemma 3.2 *If ε_k is a decreasing sequence tending to zero with $\varepsilon_{k+1} \geq \alpha\varepsilon_k$ for some $\alpha \in (0, 1)$, then*

$$d_{\rm B}(A) = \limsup_{k\to\infty} \frac{\log N(A, \varepsilon_k)}{-\log \varepsilon_k} \quad \textit{and} \quad d_{\rm LB}(A) = \liminf_{k\to\infty} \frac{\log N(A, \varepsilon_k)}{-\log \varepsilon_k}. \tag{3.4}$$

Proof Clearly the right-hand side of (3.4) is bounded by $d_{\rm B}(A)$. Given ϵ with $0 < \epsilon < 1$ let k be such that $\varepsilon_{k+1} \leq \epsilon < \varepsilon_k$; then

$$\begin{aligned} \frac{\log N(A, \epsilon)}{-\log \epsilon} &\leq \frac{\log N(A, \varepsilon_{k+1})}{-\log \varepsilon_k} \\ &= \frac{\log N(A, \varepsilon_{k+1})}{-\log \varepsilon_{k+1} + \log(\varepsilon_{k+1}/\varepsilon_k)} \\ &\leq \frac{\log N(A, \varepsilon_{k+1})}{-\log \varepsilon_{k+1} + \log \alpha} \end{aligned}$$

and so (3.4) follows. The argument for $d_{\rm LB}(A)$ is similar. □

3.2 Basic properties of the box-counting dimension

We now prove a number of properties of the box-counting dimension. In contrast to the topological (and to a lesser extent Hausdorff) dimension, the proofs are very straightforward.

Lemma 3.3 *Let (X, ϱ) be a metric space, and A and B subsets of X.*

(i) *If $A \subseteq B$ then $d_{\rm B}(A) \leq d_{\rm B}(B)$;*
(ii) *$d_{\rm B}(\overline{A}) = d_{\rm B}(A)$, where $\overline{A}$ denotes the closure of A in (X, ϱ);*

(iii) $d_{\mathrm{B}}(A \cup B) \leq \max(d_{\mathrm{B}}(A), d_{\mathrm{B}}(B))$;

(iv) *if* $f : (X, \varrho_X) \to (Y, \varrho_Y)$ *is Hölder continuous with exponent* θ,

$$\varrho_Y(f(x_1), f(x_2)) \leq C\, \varrho_X(x_1, x_2)^\theta \qquad \textit{for all} \qquad x_1, x_2 \in X$$

then

$$d_{\mathrm{B}}(f(A)) \leq d_{\mathrm{B}}(A)/\theta;$$

(v) $d_{\mathrm{H}}(A) \leq d_{\mathrm{LB}}(A) \leq d_{\mathrm{B}}(A)$; *and*

(vi) *if* $I_n = [0, 1]^n \subset \mathbb{R}^n$ *then* $d_{\mathrm{LB}}(I_n) = d_{\mathrm{box}}(I_n) = d_{\mathrm{B}}(I_n) = n$.

Proof (i) If $A \subseteq B$ then $N(A, \epsilon) \leq N(B, \epsilon)$ and the result is immediate.

(ii) Any finite cover of A by closed balls must cover $\overline{A}$, and hence $d_{\mathrm{B}}(\overline{A}) \leq d_{\mathrm{B}}(A)$. Equality follows using (i), since $A \subseteq \overline{A}$.

(iii) Clearly $N(A \cup B, \epsilon) \leq N(A, \epsilon) + N(B, \epsilon)$; the result is again immediate from the definition.

(iv) Given $d > d_{\mathrm{B}}(A)$, choose ϵ_0 sufficiently small such that $N(A, \epsilon) \leq \epsilon^{-d}$ for all $0 < \epsilon < \epsilon_0$. Cover A with no more than ϵ^{-d} balls of radius ϵ. The image of this cover under f provides a covering of $f(A)$ by sets (not necessarily closed) of diameter no larger than $C(2\epsilon)^\theta$; but these are certainly contained in closed balls of radius $2C(2\epsilon)^\theta$. So

$$N(f(A), 2C(2\epsilon)^\theta) \leq \epsilon^{-d} \quad \Rightarrow \quad N(f(A), \delta) \leq 2^d(\delta/2C)^{-d/\theta} = c\delta^{-d/\theta},$$

and hence $d_{\mathrm{B}}(f(A)) \leq d_{\mathrm{B}}(A)/\theta$.

(v) If $s > d_{\mathrm{LB}}(A)$ then there is a sequence $\epsilon_j \to 0$ such that $N(A, \epsilon_j) < \epsilon_j^{-s}$; thus

$$\mathscr{H}^s_{2\epsilon_j}(A) \leq N(A, \epsilon_j)(2\epsilon_j)^s < 2^s < \infty,$$

and hence $\mathscr{H}^s(A) < 2^s < \infty$, from which it follows that $d_{\mathrm{H}}(A) \leq s$. So $d_{\mathrm{H}}(A) \leq d_{\mathrm{LB}}(A)$; that $d_{\mathrm{LB}}(A) \leq d_{\mathrm{B}}(A)$ is immediate from the definitions.

(vi) I_n can be covered by k^n cubes of side $1/k$. The sequence $\varepsilon_k = 1/k$ satisfies the requirements of Lemma 3.2, and so $d_{\mathrm{B}}(I_n) \leq n$. We have already shown that $d_{\mathrm{H}}(I_n) = n$ (Proposition 2.8(iii)), and so $d_{\mathrm{B}}(I_n) \geq d_{\mathrm{LB}}(I_n) \geq d_{\mathrm{H}}(I_n) \geq n$. It follows that $d_{\mathrm{B}}(I_n) = d_{\mathrm{box}}(I_n) = d_{\mathrm{LB}}(I_n) = n$. □

Note that it follows immediately from parts (i) and (vi) of the above lemma that any compact subset A of $\mathbb{R}^n$ has $d_{\mathrm{B}}(A) \leq n$, and that if this A contains an open set then in fact $d_{\mathrm{B}}(A) = n$.

In Part II of this book we will be particularly interested in sets X that are subsets of Banach (or Hilbert) spaces. In this case, one can often view X simultaneously as a subset of different spaces. Since the quantity $N(X, \epsilon)$ depends on the norm in which one chooses the ϵ-balls, the box-counting dimension will vary depending on the space in which one views X as lying. In the final chapter of this book, the following simple observation will prove useful: if $\mathscr{B}_1$ and $\mathscr{B}_2$ are two Banach spaces with $X \subset \mathscr{B}_1 \subseteq \mathscr{B}_2$, then

$$\|u\|_{\mathscr{B}_2} \le c\|u\|_{\mathscr{B}_1} \qquad \Rightarrow \qquad d_{\mathrm{B}}(X;\mathscr{B}_2) \le d_{\mathrm{B}}(X;\mathscr{B}_1) \tag{3.5}$$

(the proof is immediate since $N_{\mathscr{B}_2}(X, c\epsilon) \le N_{\mathscr{B}_1}(X,\epsilon)$).

3.3 Box-counting dimension of products

In Proposition 2.10 we saw that the Hausdorff dimension obeys[2]

$$d_{\mathrm{H}}(X \times Y) \ge d_{\mathrm{H}}(X) + d_{\mathrm{H}}(Y),$$

and showed by example that this inequality could be strict. Here we show that the lower box-counting dimension behaves similarly, but that for the upper box-counting dimension the inequality is reversed.

Proposition 3.4 *Let (X, ϱ_X) and (Y, ϱ_Y) be metric spaces, and $X \times Y$ the product space equipped with the metric*

$$\varrho_\alpha((x,y),(\xi,\eta)) = [\varrho_X(x,\xi)^\alpha + \varrho_Y(y,\eta)^\alpha]^{1/\alpha} \tag{3.6}$$

for some $\alpha \in [1,\infty)$, or

$$\varrho_\infty((x,y),(\xi,\eta)) = \max(\varrho_X(x,\xi), \varrho_Y(y,\eta)). \tag{3.7}$$

Then

$$d_{\mathrm{B}}(X \times Y) \le d_{\mathrm{B}}(X) + d_{\mathrm{B}}(Y) \qquad \text{and} \qquad d_{\mathrm{LB}}(X \times Y) \ge d_{\mathrm{LB}}(X) + d_{\mathrm{LB}}(Y). \tag{3.8}$$

Consequently, if the box-counting dimensions $d_{\mathrm{box}}(X)$ and $d_{\mathrm{box}}(Y)$ are both well defined then so is $d_{\mathrm{box}}(X \times Y)$ and

$$d_{\mathrm{box}}(X \times Y) = d_{\mathrm{box}}(X) + d_{\mathrm{box}}(Y). \tag{3.9}$$

Proof For the upper bound, take $\delta_X > d_{\mathrm{B}}(X)$ and $\delta_Y > d_{\mathrm{B}}(Y)$; then there exists an $\epsilon_0 > 0$ such that

$$N(X,\epsilon) < \epsilon^{-\delta_X} \qquad \text{and} \qquad N(Y,\epsilon) < \epsilon^{-\delta_Y} \qquad 0 < \epsilon < \epsilon_0.$$

It follows that $X \times Y$ can be covered by $\epsilon^{-(\delta_X+\delta_Y)}$ balls of radius $2^{1/\alpha}\epsilon$, and hence $d_{\mathrm{B}}(X \times Y) \le d_{\mathrm{B}}(X) + d_{\mathrm{B}}(Y)$.

For the lower bound, take $s < d_{\mathrm{LB}}(X)$ and $t < d_{\mathrm{LB}}(Y)$. Then there exists an $\epsilon_0 > 0$ such that for all $\epsilon < \epsilon_0$ there are at least ϵ^{-s} disjoint balls of radius ϵ

[2] One can obtain an upper bound on the Hausdorff dimension of a product if one is prepared to involve the upper box-counting dimension: $d_{\mathrm{H}}(X \times Y) \le d_{\mathrm{H}}(X) + d_{\mathrm{B}}(Y)$, see Exercise 3.5.

with centres in X, and at least ϵ^{-t} disjoint balls of radius ϵ with centres in Y. There are certainly, therefore, more than $\epsilon^{-(s+t)}$ disjoint balls of radius $2^{1/\alpha}\epsilon$ in $X \times Y$, and hence $d_{\rm LB}(X \times Y) \geq d_{\rm LB}(X) + d_{\rm LB}(Y)$.

Finally, (3.9) follows immediately from (3.8) if $d_{\rm LB}(X) = d_{\rm B}(X) = d_{\rm B}(X)$ and $d_{\rm LB}(Y) = d_{\rm B}(Y) = d_{\rm B}(Y)$. □

Both inequalities can be strict. The example at the end of Section 2.3 can be used to show this for the lower box-counting dimension, since $n_k/m_{k+1} \to 0$ as $k \to \infty$ implies that $d_{\rm LB}(X) = d_{\rm LB}(Y) = 0$, and $d_{\rm LB}(X \times Y) \geq d_{\rm H}(X \times Y) \geq 1$ (cf. Edgar, 1998, p. 43). The construction of an example for the upper box-counting dimension seems to be more delicate: in this case one can take X and Y to be 'inhomogeneous' Cantor sets like the example used in Section 3.1, chosen in such a way that $N(X, \epsilon)$ is large (i.e. $\sim \epsilon^{-d}$ with d large) when $N(Y, \epsilon)$ is small and vice versa; see Sharples (2010) for details.

3.4 Orthogonal sequences

Now, following Ben-Artzi, Eden, Foias, and Nicolaenko (1993), we investigate the box-counting dimension of 'orthogonal sequences'. In fact we consider a class of examples that are *bona fide* orthogonal sequences in the Hilbert space ℓ^2, and behave very much like orthogonal sequences in the sequence spaces ℓ^p with $1 \leq p \leq \infty$. These examples will be used later to show that the estimates in various embedding theorems are sharp.

Let $\{e_i\}_{i=1}^{\infty}$ be the standard basis for these spaces, so that $e_i = (0, \ldots, 0, 1, 0, \ldots)$ is the sequence with 1 in the ith place and 0 in every other place.[3] Let $\{a_i\}_{i=1}^{\infty}$ be a sequence of nonzero real numbers such that $|a_n| \geq |a_{n+1}| > 0$ and $\lim_{n\to\infty} |a_n| = 0$. We consider the compact set $A = \{\alpha_1, \alpha_2, \ldots\} \cup \{0\}$, where $\alpha_i = a_i e_i$ for every $i = 1, 2, \ldots$ Since A is countable, it follows that $d_{\rm H}(A) = 0$ whatever values are chosen for the a_i.

Lemma 3.5 *In every ℓ^p, $1 \leq p \leq \infty$, the (upper) box-counting dimension of A is given by*

$$d_{\rm B}(A) = \limsup_{n\to\infty} \frac{\log n}{-\log |a_n|} \tag{3.10}$$

$$= \inf\{d : \sum_{n=1}^{\infty} |a_n|^d < \infty\}. \tag{3.11}$$

[3] This is not in fact a basis for ℓ^∞, but only for c_0, the subspace of ℓ^∞ consisting of sequences that tend to zero.

Furthermore,

$$d_{\text{LB}}(A) = \liminf_{n\to\infty} \frac{\log n}{-\log|a_n|}. \tag{3.12}$$

We will use (3.10) and (3.12) immediately to consider some simple examples. The alternative form of (3.10), (3.11), will be useful later.

Proof First we prove (3.10). Given any ϵ with $0 < \epsilon < |a_1|$, let $n = n(\epsilon)$ be the integer such that

$$|a_n| > \epsilon \geq |a_{n+1}|.$$

The set A can be covered by $n + 1$ ϵ-balls, one centred at the origin and the other n centred at $\{\alpha_1, \ldots, \alpha_n\}$. Thus

$$d_{\text{B}}(A) = \limsup_{\epsilon\to 0} \frac{\log N(X, \epsilon)}{-\log \epsilon} \leq \limsup_{\epsilon\to 0} \frac{\log n(\epsilon) + 1}{-\log|a_{n(\epsilon)}|} \leq \limsup_{n\to\infty} \frac{\log n}{-\log|a_n|}.$$

The upper bound in (3.12) follows similarly.

To prove the reverse inequality, for any n large enough that $|a_n| < 1$, let n' denote the integer $n' \geq n$ for which

$$|a_n| = |a_{n+1}| = \cdots = |a_{n'}| > |a_{n'+1}|,$$

and set $\epsilon(n) = \frac{1}{4}(|a_{n'}| + |a_{n'+1}|)$. It follows that any two elements from $\{\alpha_1, \ldots, \alpha_{n'}\}$ are at least $|a_{n'}| > 2\epsilon(n)$ apart (in any ℓ^p norm), and hence $N(A, \epsilon(n)) \geq n'$.

Since $n' \geq n$, $|a_n| = |a_{n'}|$, and $|a_{n'}| < 4\epsilon(n)$, it follows that

$$\frac{\log n}{-\log|a_n|} \leq \frac{\log n'}{-\log|a_{n'}|} \leq \frac{\log N(A, \epsilon(n))}{-\log(4\epsilon(n))},$$

and hence that

$$\limsup_{n\to\infty} \frac{\log n}{-\log|a_n|} \leq \limsup_{n\to\infty} \frac{\log N(A, \epsilon(n))}{-\log(4\epsilon(n))} \leq \limsup_{\epsilon\to 0} \frac{\log N(A, \epsilon)}{-\log(4\epsilon)} = d_{\text{B}}(A).$$

Again, the lower bound in (3.12) follows similarly.

We now show that the right-hand sides of (3.10) and (3.11), which we call d_1 and d_2 respectively, are equal. Take $d > d_2$, so that $\sum_{n=1}^{\infty} |a_n|^d = M$. Then since $|a_n|$ is nonincreasing, this implies that $n|a_n|^d \leq M$ for any n, from which it is easy to see that $d_1 \leq d$, and hence $d_1 \leq d_2$. Conversely, if $d > d_1$ then for all n sufficiently large, $\log n/(-\log|a_n|) \leq d$, and so $|a_n| \leq n^{-1/d}$. It follows that for any $d' > d$, $\sum_{n=1}^{\infty} |a_n|^{d'} < \infty$. Thus $d_2 \leq d'$ for all $d' > d_1$, and so $d_2 \leq d_1$. It follows that $d_1 = d_2$. □

As a simple application of this result, consider the set

$$H_\alpha = \{0\} \cup \{n^{-\alpha} e_n\}_{n=1}^{\infty}.$$

Then

$$d_{\rm B}(H_\alpha) = \limsup_{n\to\infty} \frac{\log n}{\alpha \log n} = \frac{1}{\alpha}. \tag{3.13}$$

More strikingly (since all these examples have zero Hausdorff dimension), the set

$$H_{\log} = \{0\} \cup \{e_n/\log n\}_{n=2}^{\infty}$$

has

$$d_{\rm B}(H_{\log}) = \limsup_{n\to\infty} \frac{\log n}{\log\log n} = +\infty.$$

By combining these two examples, one can obtain a set with $d_{\rm B}(\hat{H}) = \infty$ but $d_{\rm LB}(\hat{H}) < \infty$: the idea is to choose the coefficients $\{a_n\}$ such that $|a_{n+1}| \le |a_n|$ and there exist sequences $n_j \to \infty$ such that $a_{n_j} = 1/n_j$ and $m_j \to \infty$ such that $a_{m_j} = 1/\log m_j$. In more detail, define

$$e(x) = \lfloor \mathrm{e}^x \rfloor,$$

where $\lfloor x \rfloor$ is the greatest integer $\le x$, and then set

$$\begin{aligned} a_1 = a_2 &= 1, \\ a_3 &= 1/(\log 3), \\ a_4 = a_5 = \cdots = a_{e(4)} &= 1/4, \\ a_{e(4)+1} &= 1/\log(e(4)+1), \\ a_{e(4)+2} = \cdots = a_{e(e(4))} &= 1/(e(4)+2), \\ a_{e(e(4))+1} &= 1/\log(e(e(4))+1), \end{aligned}$$

etc. Then

$$d_{\rm B}(\hat{H}) = \limsup_{n\to\infty} \frac{\log n}{-\log |a_n|} = \infty,$$

but

$$d_{\rm LB}(\hat{H}) = \liminf_{n\to\infty} \frac{\log n}{-\log |a_n|} = 1.$$

Exercises

3.1 Show that

$$d_{\mathrm{B}}(A) = \limsup_{\epsilon \to 0} \frac{\log M(A, \epsilon)}{-\log \epsilon},$$

if $M(A, \epsilon)$ denotes:

(i) the minimum number of closed balls of radius ϵ with arbitrary centres that are required to cover A;

(ii) the largest number of disjoint balls of radius ϵ with centres in A; or

(iii) for a subset of $\mathbb{R}^n$, the number of boxes of the form

$$[m_1\epsilon, (m_1 + 1)\epsilon] \times \cdots \times [m_n\epsilon, (m_n + 1)\epsilon], \qquad m_j \in \mathbb{Z},$$

that intersect A (hence the name 'box-counting dimension').

3.2 Suppose that X is a compact subset of $\mathbb{R}^n$. Define

$$c(X) = \liminf_{\epsilon \to 0} \frac{\mathscr{L}^n(O(X, \epsilon))}{\log \epsilon},$$

where $\mathscr{L}^n$ is n-dimensional Lebesgue measure, and

$$O(X, \epsilon) = \{y \in \mathbb{R}^n : \operatorname{dist}(y, X) < \epsilon\}.$$

Show that $d_{\mathrm{B}}(X) = n - c(X)$.

3.3 Show that for any compact metric space X with $\dim(X) \leq n$ there exists a homeomorphism $h : X \to \mathbb{R}^{2n+1}$ such that

$$\dim(X) = \dim(h(X)) = d_{\mathrm{H}}(h(X)) = d_{\mathrm{LB}}(h(X))$$

and deduce that

$$\dim(X) = \inf\{d_{\mathrm{LB}}(X') : X' \text{ is homeomorphic to } X\}. \tag{3.14}$$

[Hint: let K_n consist of all mappings $f \in C(X, \mathbb{R}^{2n+1})$ such that $d_{\mathrm{LB}}(f(X)) \leq n$. Show first that for each $k \in \mathbb{N}$ the set

$$K_{n,k} = \{f \in C(X, \mathbb{R}^{2n+1}) : N_o(f(K), \epsilon) \leq \epsilon^{-n}/k \text{ for some } \epsilon > 0\}$$

is open, where $N_o(X, \epsilon)$ is the number of open balls of radius ϵ that covers X. Then use the characterisation

$$d_{\mathrm{LB}}(X) = \inf\{r : \text{ for every } \eta > 0 \text{ there exists an } \epsilon > 0 \\ \text{such that } N_o(X, \epsilon) < \eta\epsilon^{-r}\}$$

to deduce that K_n is open. Finally follow the argument of Theorem 2.12 to show that K_n is also dense and conclude the proof as there.] This result is due originally to Pontrjagin & Schnirelmann (1932); the relatively simple proof outlined here is due to Prosser (1970).

3.4 The result of the previous exercise remains true if one replaces the lower box-counting dimension by the upper box-counting dimension in (3.14). (Luukkainen (1981) pointed out that the argument of Pontrjagin & Schnirelmann (1932) can also be used to obtain this result.) Combine this with Proposition 3.4 to show that $\dim(X \times Y) \leq \dim(X) + \dim(Y)$.

3.5 Let (X, ϱ_X) and (Y, ϱ_Y) be metric spaces. For $A \subseteq X$ and $B \subseteq Y$ show that $d_{\mathrm{H}}(A \times B) \leq d_{\mathrm{H}}(A) + d_{\mathrm{B}}(B)$.

3.6 Unlike the Hausdorff dimension, the box-counting dimension is not stable under countable unions. To try to rectify this, one can introduce the *modified (upper) box-counting dimension*,

$$d_{\mathrm{MB}}(X) = \inf \left\{ \sup_i d_{\mathrm{B}}(X_i) : \; X \subset \bigcup_{i=1}^{\infty} X_i \right\}.$$

Note that $d_{\mathrm{H}}(X) \leq d_{\mathrm{MB}}(X)$ since $d_{\mathrm{H}}(X) \leq d_{\mathrm{B}}(X)$ and the Hausdorff dimension is stable under countable unions (2.4). Let X be a compact subset of a Hilbert space H, and suppose that

$$d_{\mathrm{B}}(X \cap U) = d_{\mathrm{B}}(X)$$

for all open subsets U of H that intersect X. Use the Baire Category Theorem (Theorem 1.6) to show that $d_{\mathrm{MB}}(X) = d_{\mathrm{B}}(X)$.

3.7 Set

$$\mathscr{P}^s_\delta(X) = \sup_i \{ \sum_i |B_i|^s : \; \{B_i\} \text{ are disjoint balls with centres in } X\}$$

and define

$$\mathscr{P}^s_0(X) = \lim_{\delta \to 0} \mathscr{P}^s_\delta(X).$$

To obtain a measure, define

$$\mathscr{P}^s(X) = \inf \left\{ \sum_i \mathscr{P}^s_0(X_i) : \; X \subset \bigcup_{i=1}^{\infty} X_i \right\},$$

the s-dimensional packing measure. The *packing dimension* of X is defined as

$$d_{\mathrm{P}}(X) = \inf\{s : \; \mathscr{P}^s(X) = 0\}.$$

(The definition is due to Tricot (1980).) Show that $d_{\mathrm{P}}(X) = d_{\mathrm{MB}}(X)$ (a result due to Falconer (1990)). (For more on the packing dimension see Section 5.9 in Mattila (1995), and Howroyd (1996).)

4

An embedding theorem for subsets of $\mathbb{R}^N$ in terms of the upper box-counting dimension

In this chapter we prove an embedding theorem for subsets of $\mathbb{R}^N$ in terms of the upper box-counting dimension: $X \subset \mathbb{R}^N$ can be 'nicely' embedded into $\mathbb{R}^k$ for any integer $k > 2d_B(X)$. The proof forms a model for those that follow for subsets of infinite-dimensional spaces (Theorems 6.2, 8.1, and 9.18), and motivates the definitions of 'prevalence' in Chapter 5 and of various 'thickness exponents' in Chapter 7.

The idea is to show that 'almost every' linear map $L : \mathbb{R}^N \to \mathbb{R}^k$ is one-to-one on X with $L^{-1}|_{LX}$ Hölder continuous. As remarked in the Introduction, since L is linear, $L : X \to \mathbb{R}^k$ is one-to-one if and only if $Lz = 0$ implies that $z = 0$ for $z \in X - X$, where $X - X$ is the 'difference set'

$$X - X = \{x - y : x, y \in X\}.$$

Embedding results for linear maps therefore rely essentially on properties of $X - X$ rather than on properties of X itself. For the upper box-counting dimension, however, $d_B(X - X) \le 2d_B(X)$. This follows since $d_B(X \times X) \le 2d_B(X)$ (Proposition 3.4) and $X - X$ is the image of $X \times X$ under the Lipschitz map $(x, y) \mapsto x - y$; part (iv) of Lemma 3.3 shows that such mappings cannot raise the box-counting dimension.

If X and Y are Banach spaces, we denote the space of all bounded linear maps from X into Y by $\mathscr{L}(X, Y)$, and abbreviate $\mathscr{L}(X, X)$ to $\mathscr{L}(X)$. We can view any linear map $L \in \mathscr{L}(\mathbb{R}^N, \mathbb{R}^k)$ as a collection of k linear maps $L_j : \mathbb{R}^N \to \mathbb{R}$, so that

$$Lx = (L_1 x, L_2 x, \dots, L_k x);$$

and each L_j is equivalent to taking the inner product with some $l_j \in \mathbb{R}^N$; we write l_j^* for the linear map from $\mathbb{R}^N$ into $\mathbb{R}$ given by $x \mapsto (l_j, x)$.

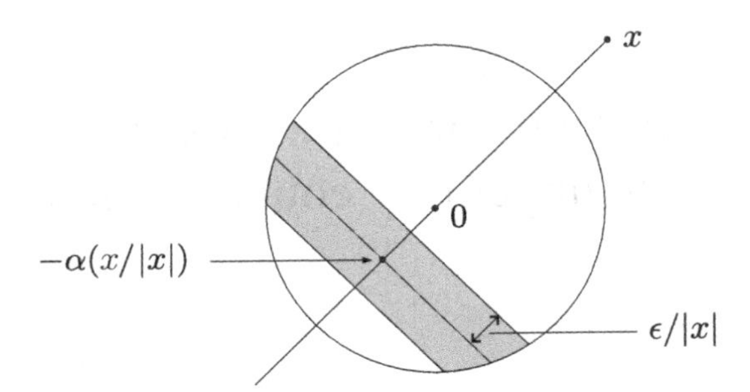

Figure 4.1 The shaded region indicates those $l \in B_N$ with $|\alpha + (l \cdot x)| \leq \epsilon$.

We will consider a restricted set E of linear maps, namely those of the form

$$E = \{(l_1^*, \ldots, l_N^*) : l_j \in B_N\},$$

where $B_N = B_N(0, 1)$ is the unit ball in $\mathbb{R}^N$. Note that any $L \in E$ has norm at most $\sqrt{N}$.

Identifying E with $(B_N)^k$, we define a probability measure μ on E to be that induced by choosing each l_j according to the uniform probability measure λ on B_N (λ is the Lebesgue measure $\mathscr{L}^N$ normalised by Ω_N), i.e. μ is the product measure $\otimes_{j=1}^k \lambda$ on $(B_N)^k$.

The following estimate lies at the heart of the proof of the embedding theorem of this chapter (Theorem 4.3).

Lemma 4.1 *For any $\alpha \in \mathbb{R}^k$ and $x \in \mathbb{R}^N$,*

$$\mu\{L \in E : |\alpha + Lx| \leq \epsilon\} \leq cN^{k/2} \left(\frac{\epsilon}{|x|} \right)^k, \tag{4.1}$$

where c is an absolute constant.

Proof Let $\alpha = (\alpha_1, \ldots, \alpha_k)$. Then

$$\begin{aligned} \mu\{L \in E : |\alpha + Lx| \leq \epsilon\} &\leq \prod_{j=1}^k \mu\{L \in E : |\alpha_j + L_j x| \leq \epsilon\} \\ &= \prod_{j=1}^k \lambda\{l \in B_N : |\alpha_j + (l \cdot x)| \leq \epsilon\}, \end{aligned}$$

using the product structure of μ. Now,

$$\lambda\{l \in B_N : |\alpha + (l \cdot x)| \leq \epsilon\} = \frac{\Omega_{N-1}}{\Omega_N} \int_{\max(-\alpha/|x|-\epsilon/|x|,-1)}^{\min(-\alpha/|x|+\epsilon/|x|,1)} (1 - r^2)^{(n-1)/2} \, \mathrm{d}r,$$

see Figure 4.1.

Since the integrand is bounded by 1 and the range of integration is no larger than $2\epsilon/|x|$,

$$\lambda\{l \in B_N : |\alpha + (l \cdot x)| \le \epsilon\} \le \frac{\Omega_{N-1}}{\Omega_N}\frac{2\epsilon}{|x|}.$$

Since $\Omega_n = \pi^{n/2}/\Gamma(n/2+1)$ and

$$\Gamma(z) = \sqrt{\frac{2\pi}{z}}\left(\frac{z}{\mathrm{e}}\right)^z(1 + O(z^{-1}))$$

(Stirling's Formula) one can deduce that

$$\lambda\{l \in B_N : |\alpha + (l \cdot x)| \le \epsilon\} \le c' N^{1/2}\frac{\epsilon}{|x|}, \tag{4.2}$$

and the inequality (4.1) now follows. □

The other key element of the proof is the Borel–Cantelli Lemma.

Lemma 4.2 (Borel–Cantelli Lemma) *Let μ be a probability measure on E, and suppose that $\{Q_j\}_{j=1}^\infty$ are subsets of E such that $\sum_{j=1}^\infty \mu(Q_j) < \infty$. Then μ-almost every element x of E lies in only finitely many of the Q_j, i.e. for each such x there exists a $j_x \in \mathbb{N}$ such that $x \notin Q_j$ for all $j \ge j_x$.*

Proof Consider

$$\mathscr{Q} = \cap_{n=1}^\infty \cup_{j=n}^\infty Q_j.$$

Then $\mathscr{Q}$ consists precisely of those $x \in E$ for which $x \in Q_j$ for infinitely many values of j. Now, for any n we must have $\mu(\mathscr{Q}) \le \mu(\cup_{j=n}^\infty Q_j) \le \sum_{j=n}^\infty \mu(Q_j)$. Since $\sum_{j=1}^\infty \mu(Q_j) < \infty$, it follows that $\sum_{j=n}^\infty \mu(Q_j) \to 0$ as $n \to \infty$, and hence $\mu(\mathscr{Q}) = 0$. □

We now put these ingredients together. Note that the following theorem only has any content if $d_{\mathrm{B}}(X) < (N-1)/2$.

Theorem 4.3 *Let X be a compact subset of $\mathbb{R}^N$. If $k > 2d_{\mathrm{B}}(X)$ then given any α with*

$$0 < \alpha < 1 - \frac{2d}{k}$$

and any linear map $L_0 \in \mathscr{L}(\mathbb{R}^N, \mathbb{R}^k)$, for μ-almost every linear map $L \in E$ there exists a $C = C_L$ such that $L' = L_0 + L$ satisfies

$$|x - y| \le C|L'x - L'y|^\alpha \qquad \text{for all} \qquad x, y \in X; \tag{4.3}$$

in particular, L' is one-to-one on X with a Hölder continuous inverse.

With $L_0 = 0$ the theorem says that μ-almost every $L' \in E$ satisfies (4.3); but the slight strengthening here is the key idea in the notion of 'prevalence' which is defined in the next chapter and allows for similar results when X is a subset of an infinite-dimensional space (Theorem 8.1). We prove a generalised version of this theorem for subsets of $\mathbb{R}^N$ using a wider class of mappings from $\mathbb{R}^N$ into $\mathbb{R}^k$ (but without the Hölder continuity of the inverse) in Lemma 14.4. The proof of the result in this form is due to Hunt & Kaloshin (1999), but earlier results along these lines, using density instead of prevalence, can be found in Ben-Artzi *et al.* (1993) and Eden *et al.* (1994).

Proof Take a fixed $L_0 \in \mathscr{L}(\mathbb{R}^N, \mathbb{R}^k)$. We try to bound the measure of linear maps L that are 'bad', i.e. do not satisfy $|(L_0 + L)z| > |z|^{1/\alpha}$ for some $z \in X - X$. To do this, we consider a collection of subproblems on a family of subsets Z_n of $X - X$ that are bounded away from the origin: define

$$Z_n = \{z \in X - X : |z| \geq 2^{-n}\}$$

and set

$$Q_n = \{L \in E : |(L_0 + L)z| \leq 2^{-n/\alpha} \text{ for some } z \in Z_n\}.$$

This Q_n is essentially the set of 'bad' linear maps for which (4.3) (with $C = 1$) does not hold for some (x, y) with $|x - y| \geq 2^{-n}$.

We now use the fact that $d_{\mathrm{B}}(X - X) \leq 2d_{\mathrm{B}}(X)$. Choose and fix $d > d_{\mathrm{B}}(X)$; then $Z_n \subset X - X$ can be covered by a collection of no more than $N := 2^{2nd/\alpha}$ balls of radius $2^{-n/\alpha}$, $\{B(z_j, 2^{-n/\alpha})\}$, whose centres z_j lie in Z_n.

Let $Y_j = Z_n \cap B(z_j, 2^{-n/\alpha})$. Now note that if

$$|(L_0 + L)z_j| > 2^{-n/\alpha}(1 + 2\sqrt{k} + 2\|L_0\|)$$

then

$$|(L + L_0)z| > 2^{-n/\alpha} \quad \text{for every } z \in Y_j.$$

Define $M := 1 + 2\sqrt{k} + 2\|L_0\|$. It follows that if $|(L_0 + L)z| \leq 2^{-n/\alpha}$, then $L_0 + L$ must map z_j close to the origin,

$$|(L_0 + L)z_j| \leq 2^{-n/\alpha}M.$$

See Figure 4.2.

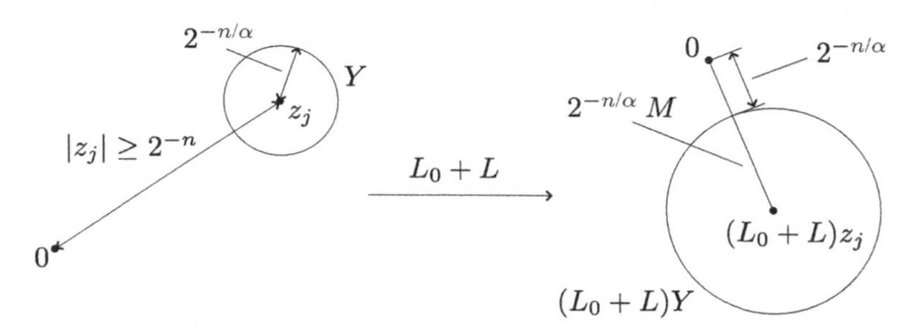

Figure 4.2 If $|(L_0 + L)z_j| > 2^{-n/\alpha}M$ then $|(L_0 + L)z| > 2^{-n/\alpha}$ for every $z \in Y$.

Since $|z_j| \geq 2^{-n}$, it follows using (4.1) that

$$\begin{aligned}
&\mu\{L \in E : |(L_0 + L)z| \leq 2^{-n/\alpha} \text{ for some } z \in Z_n \cap B(z_j, 2^{-n/\alpha})\} \\
&\quad \leq \mu\{L \in E : |(L_0 + L)z_j| \leq 2^{-n/\alpha} M\} \\
&\quad = \mu\{L \in E : |(L_0 z_j) + L z_j| \leq 2^{-n/\alpha} M\} \\
&\quad \leq C_{N,k} \left(\frac{2^{-n/\alpha} M}{|z_j|} \right)^k \leq C'_{N,k,L_0} \left(\frac{2^{-n/\alpha}}{2^{-n}} \right)^k \\
&\quad = C'_{n,k,L_0} 2^{nk(1-(1/\alpha))}.
\end{aligned}$$

Thus the total measure of Q_n, i.e. those maps for which things fail for some $z \in Z_n$, is bounded by

$$\mu(Q_n) \leq 2^{2nd/\alpha} \cdot C'_{N,k,L_0} 2^{nk(1-(1/\alpha))} = C'_{N,k,L_0} \cdot 2^{[k-(k-2d)/\alpha]n}.$$

To apply the Borel–Cantelli Lemma (Lemma 4.2) we require that $\sum_{n=1}^{\infty} \mu(Q_n) < \infty$, and this is ensured if

$$k - \frac{(k-2d)}{\alpha} < 0.$$

This means that we must take $k > 2d$, and then $\alpha < 1 - (2d/k)$ as in the statement of the theorem. Thus μ-almost every L lies in only a finite number of the Q_j: for such an L, there exists a j_L such that $L \notin Q_j$ for all $j \geq j_L$, i.e.

$$|z| \geq 2^{-j} \quad \Rightarrow \quad |(L_0 + L)z| \geq 2^{-j/\alpha} \qquad \text{for all} \qquad j \geq j_L.$$

Then, if $X - X \subset B(0, R)$, for $|z| > 2^{-j_L}$

$$|(L_0 + L)z| \geq 2^{-j_L/\alpha} \geq \frac{2^{-j_L/\alpha}}{R^{1/\alpha}} |z|^{1/\alpha},$$

while if $2^{-(j+1)} < |z| \le 2^{-j}$ with $j \ge j_L$ then

$$|(L_0 + L)z| \ge 2^{-(j+1)/\alpha} \ge 2^{-1/\alpha}|z|^{1/\alpha},$$

from which it follows that

$$|(L_0 + L)z| \ge \max\left[2^{-1/\alpha}, \frac{2^{-j_L/\alpha}}{R^{1/\alpha}}\right]|z|^{1/\alpha},$$

which implies (4.3). □

In order to prove a similar result for a subset of an infinite-dimensional Hilbert space H (in fact we will also cover the case of a general Banach space) there are a number of ingredients that we need to adapt.

Firstly, we require a notion of what might be meant by 'almost every' linear map from H into $\mathbb{R}^k$. This is provided by the concept of 'prevalence', discussed in the next chapter; essentially we have to find an analogue of the distinguished space E of linear maps, and define an appropriate probability measure μ on E. A set S of linear maps is then prevalent if for every $L_0 \in \mathscr{L}(H, \mathbb{R}^k)$, $L_0 + L \in S$ for μ-almost every $L \in E$.

Secondly, given a suitable space E and measure μ, we will require a version of the key inequality (4.1),

$$\mu\{L \in E : |\alpha + Lx| \le \epsilon\} \le cN^{k/2}\left(\frac{\epsilon}{|x|}\right)^k, \tag{4.4}$$

that holds in this more general setting. In a Hilbert space H we can construct the space E based on a sequence V_j of finite-dimensional subspaces of H, in such a way that the estimate

$$\mu\{L \in E : |\alpha + Lx| \le \epsilon\} \le c\,(\dim V_j)^{k/2}\left(\frac{\epsilon}{\|P_j x\|}\right)^k \tag{4.5}$$

holds, where P_j is the orthogonal projection onto V_j. We provide this, and an equivalent result in Banach spaces, once we have introduced the notion of prevalence in the following chapter.

Finally, inequality (4.4) was used in the proof of the theorem above with $\epsilon \sim 2^{-n/\alpha}$ and $|x| \ge 2^{-n}$. To be able to make use of a similar argument using (4.5), when $\|x\| \ge 2^{-n}$ we would like (ideally) to have $\|P_j x\| \ge c2^{-n}$. This is possible if X lies sufficiently close (within $2^{-n}/3$, say) to the space V_j. Since the dimension of V_j occurs in (4.5), we will require some control over how $\dim V_j$ grows as $\mathrm{dist}(X, V_j)$ decreases. This is provided by the thickness exponent (and variants), which are discussed in Chapter 7.

5
Prevalence, probe spaces, and a crucial inequality

The term 'prevalence' was coined by Hunt *et al.* (1992), for a generalisation of the notion of 'almost every' that is appropriate for infinite-dimensional spaces. Essentially the same definition was used earlier by Christensen (1973), although for him a set was prevalent if its complement was a Haar null set; we adopt here the more recent and more descriptive terminology. A nice review of the theory of prevalence is given by Ott & Yorke (2005). We only develop the theory here as far as we will need it in what follows; more details can be found in the above papers and in Benyamini & Lindenstrauss (2000, Chapter 6).

Once we have introduced prevalence, we show how the idea can be adapted to treat certain classes of linear maps from infinite-dimensional spaces into finite-dimensional Euclidean spaces (Section 5.2), and then prove a generalisation of the inequality (4.1) that is a key element of the subsequent embedding proofs.

5.1 Prevalence

Let V be a normed linear space. First we define what it means for a subset of V to be 'shy', the equivalent in this setting of 'having measure zero'; the complement of a shy set is said to be 'prevalent'.

Definition 5.1 A Borel set $S \subset V$ is *shy* if there exists a compactly supported probability measure[1] μ on V such that

$$\mu(S + v) = 0 \qquad \text{for every} \quad v \in V. \tag{5.1}$$

More generally, a set is shy if it is contained in a shy Borel set.

[1] Hunt *et al.* (1992) in fact make what initially appears to be a weaker definition: there need only exist some measure μ such that $0 < \mu(U) < \infty$ for some compact set U, for which (5.1) holds. They then, however, make the observation that given such a measure one can always take instead an appropriately weighted restriction of μ to U to obtain a compactly supported probability measure for which (5.1) still holds.

It is easy to show that in $\mathbb{R}^n$ a set is shy if and only if it has measure zero.

Lemma 5.2 *If $S \subset \mathbb{R}^n$ then S is shy if and only if its Lebesgue measure is zero.*

Proof Since subsets of Borel sets with Lebesgue measure zero also have Lebesgue measure zero, and the same is true of 'shyness', we need only consider Borel sets. If a Borel set S has Lebesgue measure zero then one can take μ to be Lebesgue measure on the unit ball in $\mathbb{R}^n$ (weighted by the inverse of the volume of the ball), and clearly $\mu(S+v)=0$ for every $v \in \mathbb{R}^n$.

Conversely, let S be a Borel set and suppose that there exists a compactly supported probability measure μ such that (5.1) holds for every $v \in \mathbb{R}^n$; let ν be Lebesgue measure. Then by the Tonelli Theorem

$$0=\int_{\mathbb{R}^n}\mu(S-y)\,\mathrm{d}\nu(y)=\int_{\mathbb{R}^n}\nu(S-x)\,\mathrm{d}\mu(x)=\nu(S)\mu(\mathbb{R}^n)=\nu(S),$$

and so S has Lebesgue measure zero. □

A set is *prevalent* if its complement is shy. For a more intuitive version of the definition of prevalence, one can think of $E=\mathrm{supp}(\mu)$ as a 'probe space' of allowable perturbations: then S is prevalent if for every $v \in V$, $v+e \in S$ for μ-almost every $e \in E$.

In this form it is clear that if S is prevalent then S is dense: given any $\epsilon>0$, since E is compact it can be covered by a finite number of balls of radius ϵ. At least one of these balls, $B(x,\epsilon)$, has positive μ-measure. So for any $v \in V$, $v+B(0,\epsilon)=v-x+B(x,\epsilon)$ contains a point of S.

We now show that the union of a finite number of shy sets is shy; this requires some proof since each set may be 'shy' with respect to a different measure.

Lemma 5.3 *The union of a finite number of shy sets is shy.*

Proof We show that the union of two shy sets is shy, and the result then follows by induction. To this end, given two shy sets S' and T', find shy Borel sets S and T that contain them, with corresponding probability measures μ and ν.

Let $\mu\times\nu$ be the product measure on $V\times V$, and for a Borel set $S\subset V$ define

$$S_\Sigma=\{(x,y)\in V\times V:\ x+y\in S\}.$$

Then S_Σ is a Borel subset of $V\times V$, and we define

$$\mu*\nu(S)=(\mu\times\nu)(S_\Sigma).$$

Since

$$\mu * \nu(S) = \int_V \mu(S - y)\,\mathrm{d}\nu(y) = \int_V \nu(S - x)\,\mathrm{d}\mu(x),$$

it follows that

$$\begin{aligned}\mu * \nu([S \cup T] + v) &\le \mu * \nu(S + v) + \mu * \nu(T) \\ &= \int_V \mu(S + v - y)\,\mathrm{d}\nu(y) + \int_V \nu(T - x)\,\mathrm{d}\mu(x) = 0\end{aligned}$$

for all $v \in V$, and so $S \cup T$ is shy. □

Corollary 5.4 *The intersection of a finite number of prevalent sets is prevalent.*

With a little more work one can show that the countable union of shy sets is shy, and so the countable intersection of prevalent sets is prevalent. We will not require this (potentially powerful) result in what follows; a proof can be found in Hunt *et al.* (1992, Fact 3″), Ott & Yorke (2005, Axiom 3), or Benyamini & Lindenstrauss (2000, Proposition 6.3).

5.2 Measures based on sequences of linear subspaces

In the theorems that follow that give embeddings of finite-dimensional subsets of infinite-dimensional spaces into Euclidean spaces, the construction of an appropriate probe space E and the associated measure μ, tailored to the set (and to the particular definition of dimension being considered) is critical. While the exact choice will vary, all the constructions we will use fit into the following general framework, which gives a compactly supported probability measure on the space $\mathscr{L}(\mathscr{B}, \mathbb{R}^k)$ of all bounded linear maps from some Banach space $\mathscr{B}$ into $\mathbb{R}^k$. The basic construction, along with the proof of Lemma 5.6 and the key ideas behind the proof of Lemma 5.9, is due to Hunt & Kaloshin (1999).

In the case of a Hilbert space the construction is slightly more straightforward, and obtaining bounds on the measure of linear maps in E that map a given x close to the origin is elementary (essentially we have already made the required estimates in Lemma 4.1). We treat this case first, before considering the construction in Banach spaces.

5.2.1 The probe set and its measure in a Hilbert space

Let H be a real Hilbert space, and $\mathscr{V} = \{V_j\}_{j=1}^{\infty}$ a sequence of finite-dimensional linear subspaces of H. Denote by d_j the dimension of V_j, and let S_j be the unit ball in V_j; using an orthonormal basis for V_j we can identify S_j with B_{d_j}, the unit ball in $\mathbb{R}^{d_j}$.

Given any $l \in H$, we denote by l^* the element of H^* (the dual of H) given by $l^*(x) = (l, x)$. For a fixed $\gamma > 0$ define the probe space $E_\gamma(\mathscr{V})$ (we will call this space E for short) as the collection of all maps $L : H \to \mathbb{R}^k$ given by

$$E_\gamma(\mathscr{V}) = \left\{ L = (l_1, \ldots, l_k) : \ l_n = \left(\sum_{i=1}^{\infty} i^{-\gamma} \phi_{n,i} \right)^*, \quad \phi_{n,i} \in S_i \right\}.$$

Clearly $E = E_0^k$, where

$$E_0 = \left\{ \left(\sum_{i=1}^{\infty} i^{-\gamma} \phi_i \right)^* : \ \phi_i \in S_i \right\}.$$

The factor $i^{-\gamma}$ in the expression for l_n is there to ensure convergence of the sum. In general, we need $\gamma > 1$ (and convergence then follows using the triangle inequality), but in the particular case that the spaces V_j are orthogonal it suffices to take $\gamma > 1/2$ (we will only make use of this observation in the proof of Theorem 9.18, an embedding result involving the Assouad dimension). It is straightforward to show that E is compact (see Exercise 5.2).

To define a measure on E, we first define a probability measure λ_i on each S_i by identifying S_i with B_{d_i}, and using the uniform probability measure on B_{d_i}. Then each $\phi_{n,i}$ is chosen at random and independently using the measure λ_i on S_i. To formalise this, we consider the product space

$$\mathbb{E} = \mathbb{E}_0^k := \left(\prod_{i=1}^{\infty} S_i \right)^k,$$

and define a measure μ on E to be that obtained from k copies of the product measure

$$\mu_0 := \bigotimes_{i=1}^{\infty} \lambda_i$$

defined on $\mathbb{E}_0$ (so that μ_0^k is defined on $\mathbb{E}$).

Our aim, given $f \in \mathscr{L}(H, \mathbb{R}^k)$ and $x \in H$, is to find a bound on

$$\mu\{ L \in E : \ |(f + L)(x)| < \epsilon \}.$$

We have essentially already obtained the following simplified version of this estimate (from which the bound we require follows fairly easily) in Lemma 4.1.

Lemma 5.5 *If $\alpha \in \mathbb{R}$ and $x \in H$ then*

$$\lambda_j\{\phi \in S_j : |\alpha + (\phi, x)| < \epsilon\} \leq c d_j^{1/2} \left(\frac{\epsilon}{\|P_j x\|}\right), \tag{5.2}$$

where P_j is the orthogonal projection onto V_j, and c is a constant which does not depend on α or j.

Proof We identify V_j with $\mathbb{R}^{d_j}$ and S_j with B_{d_j} in the obvious way. Noting that for $v \in S_j$ we have $(v, x) = (v, P_j x)$, the estimate follows immediately from (4.2). □

Given the result of the previous lemma, the following key estimate is relatively straightforward.

Lemma 5.6 *If $x \in H$ and $f \in \mathscr{L}(H, \mathbb{R}^k)$ then for every j,*

$$\mu\{L \in E : |(L + f)(x)| < \epsilon\} \leq c \left(j^\gamma d_j^{1/2} \frac{\epsilon}{\|P_j x\|}\right)^k, \tag{5.3}$$

where c is a constant independent of j and f, and P_j is the orthogonal projection onto V_j.

Proof We wish to bound

$$\begin{aligned} &\mu\{L \in E : |(f + L)(x)| < \epsilon\} \\ &\quad \leq \mu\{L = (l_1, \ldots, l_k) \in E : |(f_n + l_n)(x)| < \epsilon \text{ for each } n = 1, \ldots, k\} \\ &\quad = \prod_{n=1}^{k} \mu_0\{l \in E_0 : |(f_n + l)(x)| < \epsilon\}. \end{aligned}$$

So we take an $f_0 \in H^*$ and consider

$$\begin{aligned} &\left[\bigotimes_{i=1}^{\infty} \lambda_i\right] \left\{ \{\phi_i\}_{i=1}^{\infty} \in \mathbb{E}_0 : \left| f_0(x) + \sum_{i=1}^{\infty} i^{-\gamma}(\phi_i, x) \right| < \epsilon \right\} \\ &\quad = \left[\bigotimes_{i=1}^{\infty} \lambda_i\right] \left\{ \{\phi_i\}_{i=1}^{\infty} \in \mathbb{E}_0 : \left| \left[f_0(x) + \sum_{i \neq j}^{\infty} i^{-\gamma}(\phi_i, x) \right] + j^{-\gamma}(\phi_j, x) \right| < \epsilon \right\}. \end{aligned}$$

Lemma 5.5 shows that for $\alpha = f_0(x) + \sum_{i \neq j}^{\infty} i^{-\gamma}(\phi_i, x)$ fixed, the bound on

$$\lambda_j\{\phi \in S_j : |\alpha + j^{-\gamma}(\phi, x)| < \epsilon\}$$

is independent of α. It follows from the product structure of the measure $\otimes_{j=1}^{\infty}\lambda_j$ that

$$\mu_0\{l \in E_0 : |(f_n + l)(x)| < \epsilon\} \leq \lambda_j\{\phi \in S_j : |j^{-\gamma}(\phi, x)| < \epsilon\},$$

and the inequality (5.3) now follows from the estimate (5.2). □

In the proof of Theorem 9.18 we will require a more refined result. We specialise to the case in which the $\{V_j\}$ are mutually orthogonal, and $\dim(V_j) \leq d$ for every j. Rather than using S_j, the unit ball in V_j, in our construction of E, we instead use a 'unit cube' C_j, where

$$C_j = \{u \in V_j : |(u, e_{j,i})| \leq \tfrac{1}{2},\ i = 1 \ldots, j\},$$

with $\{e_{j,i}\}_{i=1}^{d_j}$ an orthonormal basis for V_j. The measure on C_j is now induced by Lebesgue measure on $I_{d_j} := [-\frac{1}{2}, \frac{1}{2}]^{d_j}$. Since any element of C_j has norm bounded by $\sqrt{d}$, we can use the orthogonality of the $\{V_j\}$ to allow any $\gamma > 1/2$ in the definition of E.

We will require the following result of Ball (1986) about the volume of $(d-1)$-dimensional slices through the unit cube in $\mathbb{R}^d$. The key point is that the upper bound (which is sharp) does not depend on the dimension d. (Hensley (1979) proved a similar result but with the upper bound 5.)

Theorem 5.7 *Let $I_d = [-\frac{1}{2}, \frac{1}{2}]^d$ be the unit ball in $\mathbb{R}^d$, and let S be a codimension 1 subspace in $\mathbb{R}^d$ with unit normal a. Then for any $r \in \mathbb{R}$*

$$\mathscr{L}^{d-1}((S + ra) \cap I_d) \leq \sqrt{2}.$$

Given this, we can prove the following bound.

Lemma 5.8 *In the situation described above, given any $x \in H$ and any $f \in \mathscr{L}(H, \mathbb{R}^k)$, for any j*

$$\mu\{L \in E : |(L + f)(x)| < \epsilon\} \leq c\left(j^{\gamma} d^{1/2}\frac{\epsilon}{\|\Pi_j x\|}\right)^k, \tag{5.4}$$

where c is a constant independent of j and f, and Π_j is the orthogonal projection onto $V_1 \oplus V_2 \oplus \cdots \oplus V_j$.

Proof Arguing as in the proof of Lemma 5.6, the left-hand side of (5.4) is bounded by

$$\left[\bigotimes_{j=1}^{n}\lambda_j\right]\{(\phi_1, \ldots, \phi_n) \in \prod_{j=1}^{n} C_j : \left|\sum_{j=1}^{n} j^{-\gamma}\phi_j^*(\Pi_j x)\right| < \epsilon\}.$$

As in Lemma 5.3, the estimate now depends on an entirely finite-dimensional problem. Indeed, each $V_j \simeq \mathbb{R}^{d_j}$, and C_j (the 'unit cube' in V_j) is isomorphic to I_{d_j}. Set $D = \sum_{j=1}^{n} d_j$. The vector $(P_1 x, \dots, P_n x)$ corresponds to a vector $a = (a_1, \dots, a_n) \in \mathbb{R}^D$; if we set

$$a' = (a_1, 2^{-s} a_2, \cdots, n^{-s} a_n) \qquad \text{and} \qquad \hat{a} = a'/|a'|$$

and let μ denote the uniform probability measure on I_D (i.e. Lebesgue measure), the problem is to bound, for any $y \in \mathbb{R}$,

$$\begin{aligned}\mu\{x \in I_D : |y + (x \cdot a')| \le \epsilon\} &= \frac{1}{|a'|} \mu\{x \in I_D : |y + (x \cdot \hat{a})| \le \epsilon\} \\ &\le \frac{n^s}{|a|} \mu\{x \in I_D : |y + (x \cdot \hat{a})| \le \epsilon\},\end{aligned}$$

where $\hat{a} = a'/|a'|$. The result is now a consequence of Theorem 5.7, since

$$\mu\{x \in I_D : |y + (x \cdot \hat{a})| \le \epsilon\} \le 2\epsilon |(S_{\hat{a}} - y\hat{a}) \cap I_D| \le 2\epsilon\sqrt{2},$$

where $S_{\hat{a}}$ is the hyperplane through the origin with normal $\hat{a}$. □

5.2.2 The probe set and its measure in a Banach space

Kakutani (1939) showed that if there is an linear isometry from the dual of each finite-dimensional subspace of $\mathscr{B}$ onto some linear subspace of $\mathscr{B}^*$, then $\mathscr{B}$ must be a Hilbert space. This means that we cannot extend directly the construction of the previous section – where we associated elements of S_j to elements of H^* via the Riesz mapping $x \mapsto (x, \cdot)$ – to the Banach space case. In order to circumvent this problem, we can use a similar construction, but one that begins with a sequence of subspaces of $\mathscr{B}^*$ rather than of $\mathscr{B}$.

So let $\mathscr{V} = \{V_j\}_{j=1}^{\infty}$ be a sequence of finite-dimensional linear subspaces of $\mathscr{B}^*$, let d_j denote the dimension of V_j, and let S_j be the unit ball in V_j. For a fixed $\gamma > 0$ define the probe space $E_\gamma(\mathscr{V})$ to be the collection of all maps $L : \mathscr{B} \to \mathbb{R}^k$ given by

$$E_\gamma(\mathscr{V}) = \left\{ L = (L_1, \dots, L_k) : L_n = \sum_{j=1}^{\infty} j^{-\gamma} \phi_{n,j}, \quad \phi_{n,j} \in S_j \right\}.$$

To define a measure on E, first choose a basis for V_j, so that by means of the coordinate representation with respect to this basis one can identify S_j with a symmetric convex set $U_j \subset \mathbb{R}^{d_j}$ (recall that $\dim V_j = d_j$). The uniform probability measure on U_j (Lebesgue measure normalised by the volume of U_j) induces the probability measure λ_j on S_j; we now proceed as before, choosing each $\phi_{n,j}$ independently and at random according to the measure λ_j.

Of course, one could just as well define the probe space this way in the Hilbert space setting; in terms of the construction outlined in Section 5.2.1, the subspaces V_j of H are simply replaced by the subspaces V_j^* obtained by the isometry $x \mapsto (x, \cdot)$; the unit ball S_j in V_j corresponds to the unit ball in V_j^* under the same mapping, and $U_j = B_{d_j}(0, 1)$.

The proof of the Banach space version of Lemma 5.5 is significantly more involved than that for the Hilbert space case.

Lemma 5.9 *If $\alpha \in \mathbb{R}$ and $x \in \mathscr{B}$ then*

$$\lambda_j\{\phi \in S_j : |\alpha + \phi(x)| < \epsilon\} \leq d_j \left(\frac{\epsilon}{|g(x)|} \right), \tag{5.5}$$

for any $g \in S_j$.

Proof Write ρ for the left-hand side of (5.5). If $g(x) = 0$ then the inequality is trivially true. So assume that $g(x) \neq 0$, and let P be the subspace of $\mathscr{B}^*$ that annihilates x.

If h is any other element of S_j with $h(x) \neq 0$ then since

$$[g(x)h - h(x)g](x) = 0,$$

it follows that $g(x)h = h(x)g + p$ for some $p \in P$. One can therefore write any element of S_j in the form $p + rg$ for some $p \in P$ and $r \in \mathbb{R}$. That this expansion is unique can be seen easily by applying both sides of $p_1 + r_1 g = p_2 + r_2 g$ to x.

Now, ρ is bounded above by the probability that $\phi \in S_j$ lies between

$$\left(-\frac{r}{g(x)} - \frac{\epsilon}{|g(x)|} \right) g + P \qquad \text{and} \qquad \left(-\frac{r}{g(x)} + \frac{\epsilon}{|g(x)|} \right) g + P.$$

If P is represented by the hyperplane Π in $\mathbb{R}^{d_j}$, and g by the vector γ, then by definition this is the fraction of the measure of U_j that lies between

$$(-\beta - \epsilon|g(x)|^{-1})\gamma + \Pi \qquad \text{and} \qquad (-\beta + \epsilon|g(x)|^{-1})\gamma + \Pi.$$

Now consider the intersections of U_j with translates of Π: for $s \in \mathbb{R}$ set

$$U_j \cap (\Pi + s\gamma) = K_s. \tag{5.6}$$

It follows from the Brunn–Minkowski Inequality (see Exercise 5.3) that $\mathscr{L}^{d_j - 1}(K_s)$ attains its maximal value Λ when $s = 0$, i.e. on the 'slice' through the origin.

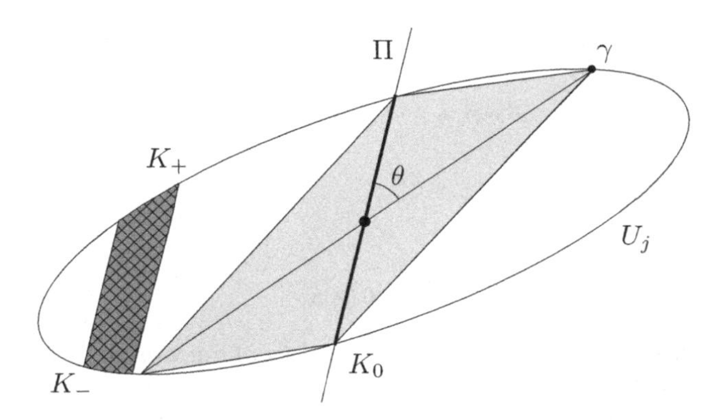

Figure 5.1 The hatched area, between $K_+ = K_{-\beta+(\epsilon/|g(x)|)}$ and $K_- = K_{-\beta-(\epsilon/|g(x)|)}$ indicates those elements of U_j corresponding to some $\phi \in S_j$ for which $|\alpha + \phi(x)| < \epsilon$. The dark line represents K_0, the 'slice' with maximal $(d_j - 1)$-volume Λ. The lightly shaded cone provides a lower bound on the measure of U_j.

If θ denotes the (smallest) angle that γ makes with Π then

$$\rho\, \mathscr{L}^{d_j}(U_j) \leq \Lambda|\gamma| \left(\frac{2\epsilon}{|g(x)|}\right) \sin\theta.$$

Since U_j is convex it contains the cone with base $K_0 = U_j \cap \Pi$ and vertex γ, along with its mirror image (i.e. the cone with base K_0 and vertex $-\gamma$). Thus

$$\mathscr{L}^{d_j}(U_j) \geq \frac{2\Lambda|\gamma|\sin\theta}{d_j},$$

and so

$$\rho \leq d_j \left(\frac{\epsilon}{|g(x)|}\right) \tag{5.7}$$

and (5.5) follows. See Figure 5.1. □

Note that one cannot improve significantly on the argument leading to (5.7), since the 'double cone'

$$U = \{(x_1, \ldots, x_n) : \ |(x_1, \ldots, x_{n-1})| \leq 1 - |x_n|,\ |x_n| \leq 1\}$$

is a convex symmetric subset of $\mathbb{R}^n$ whose volume is Ω_{n-1}/n, and hence the ratio of the largest $(n-1)$-dimensional 'slice' through the origin (Ω_{n-1}) to the volume is precisely n.

We now follow the argument of Lemma 5.6, using the estimate (5.5), to obtain the Banach-space version of (5.3).

Lemma 5.10 *If $x \in \mathscr{B}$ and $f \in \mathscr{L}(\mathscr{B}, \mathbb{R}^k)$ then for every $j \in \mathbb{N}$*

$$\mu\{L \in E : |(f + L)(x)| < \epsilon\} \leq \left(j^\gamma d_j \frac{\epsilon}{|g(x)|}\right)^k \tag{5.8}$$

for any $g \in S_j$.

Exercises

5.1 Show that $\int_0^1 f(x)\,\mathrm{d}x \neq 0$ for a prevalent set of functions in $f \in L^1(0, 1)$.

5.2 Show that $E_\gamma(\mathscr{V})$ is a compact subset of $\mathscr{L}(H, \mathbb{R}^k)$.

5.3 The Brunn–Minkowski Inequality (see Gardner (2002), for example) says that if L and M are two convex subsets of $\mathbb{R}^n$ then

$$\mathscr{L}^n((1-t)L + tM)^{1/n} \geq (1-t)\mathscr{L}^n(L)^{1/n} + t\mathscr{L}^n(M)^{1/n}$$

for $t \in [0, 1]$. Use this to show that the map $s \mapsto \mathscr{L}^{d_j - 1}(K_s)^{1/(d_j-1)}$ is concave, where K_s is defined in (5.6), and deduce that $\mathscr{L}^{d_j-1}(K_s)$ attains its maximal value when $s = 0$.

6
Embedding sets with $d_{\mathrm{H}}(X - X)$ finite

We now give the first application of the constructions of the previous chapter to prove a 'prevalent' version of a result first due to Mañé (1981). He showed that if X is a subset of a Banach space $\mathscr{B}$ and $d_{\mathrm{H}}(X - X) < k$, then a residual subset of the space of projections onto any subspace of dimension at least k are injective on X.

We show here that in general no linear embedding into any $\mathbb{R}^k$ is possible if we only assume that $d_{\mathrm{H}}(X)$ is finite (Section 6.1). If we want an embedding theorem for such sets, we must fall back on Theorem 1.12 which guarantees the existence of generic embeddings of sets with finite covering dimension (we can apply this result since $\dim(X) \leq d_{\mathrm{H}}(X)$ by Theorem 2.11).

While we prove in Theorem 6.2 the existence of a prevalent set of linear embeddings into $\mathbb{R}^k$ when $d_{\mathrm{H}}(X - X) < k$, we will see that even with this assumption one cannot guarantee any particular degree of continuity for the inverse of the linear mapping that provides the embedding (Section 6.3).

In this chapter and those that follow, we will often wish to show that certain embedding results are sharp, in the sense that the information we obtain on the modulus of continuity for the inverse of the embedding map cannot be improved. In this context, the following decomposition lemma, which allows us to reduce the analysis of general linear maps to the analysis of orthogonal projections, is extremely useful.

Lemma 6.1 *Let H be a Hilbert space and suppose that $L : H \to \mathbb{R}^k$ is a linear map with $L(H) = \mathbb{R}^k$. Then $U = (\ker L)^\perp$ has dimension k, and L can be decomposed uniquely as MP, where P is the orthogonal projection onto U and $M : U \to \mathbb{R}^k$ is an invertible linear map.*

Proof Let $U = (\ker L)^\perp$ and suppose that there exist $m > k$ linearly independent elements $\{x_j\}_{j=1}^m$ of U for which $Lx_j \neq 0$. Then $\{Lx_j\}_{j=1}^m$ are elements of $\mathbb{R}^k$; since $m > k$ at least one of the $\{Lx_j\}$ can be written as a linear combination of the others:

$$Lx_i = \sum_{j \neq i} c_j (Lx_j).$$

It follows that

$$L\Big(x_i - \sum_{j \neq i} c_j x_j\Big) = 0,$$

and hence

$$x_i - \sum_{j \neq i} c_j x_j \in \ker L \cap (\ker L)^\perp = \{0\}.$$

Thus $x_i = \sum_{j \neq i} c_j x_j$ and the $\{x_j\}_{j=1}^m$ are not linearly independent, which contradicts the definition of U.

Let P denote the orthogonal projection onto U, and M the restriction of L to U. Take $x \in H$, and decompose $x = u + v$, where $u \in U$ and $v \in \ker L$. Note that this decomposition is unique. Clearly $Lx = Lu = Mu = M(Px)$. It remains to show that M is invertible. This is clear since $\dim U = \dim \mathbb{R}^k = k$ and M is linear. □

We state a Banach space version of this result in Lemma 8.2.

6.1 No linear embedding is possible when $d_{\rm H}(X)$ is finite

It is not possible to prove a result guaranteeing the existence of injective linear maps (or projections) for sets with finite Hausdorff dimension. Kan (in the appendix to the paper of Sauer *et al.* (1991)) gave the following construction of a compact subset K_m of $\mathbb{R}^m$ such that no proper projection of $\mathbb{R}^m$ is injective on K_m. Once such a set K_m is constructed, we will follow Ben-Artzi *et al.* (1993) to find a subset K of an infinite-dimensional Hilbert space that has zero Hausdorff dimension but for which no linear map into any finite-dimensional Euclidean space can be injective.

For each m, the set K_m is formed by the union of two sets A and B, with the property that the images of A and B under any proper projection of $\mathbb{R}^m$ must intersect.

First, let C be the Cantor set formed of all x whose binary expansion $x = x^1 x^2 x^3 \cdots$ has $x^l = 0$ for every $l \in (M_{2k}, M_{2k+1}]$, or $x^l = 1$ for every $l \in (M_{2k}, M_{2k+1}]$, where the sequence M_k is chosen so that

$$0 = M_0 < M_1 < M_2 < \cdots \qquad \text{and} \qquad \lim_{k\to\infty} \frac{M_{k+1}}{M_k} = +\infty$$

(e.g. $M_k = 0$ and $M_k = 2^{k^2}$ for $k \geq 1$). The set C can be covered by 2^{r_k} intervals of length $2^{-M_{2k+1}}$, where

$$r_k = k + \sum_{j=1}^{k}(M_{2j} - M_{2j-1}) = k + M_{2k} - 1.$$

The set A is given as the union of m sets A_j, each lying on a face of the unit m-cube. A_j consists of points $a = (a_1, \ldots, a_m)$ with $a_j = 0$ and a_i for $i \neq j$ an element of the Cantor set C constructed above.

Since for $i \neq j$, the one-dimensional orthogonal projection of A_j onto the ith coordinate axis is precisely C, it follows that A_j can be covered by $2^{(m-1)r_k}$ cubes whose edges have length $2^{-M_{2k+1}}$. It is easy to see that for any $s > 0$,

$$2^{(m-1)r_k}[2^{-M_{2k+1}}]^s \to 0$$

as $k \to \infty$, and hence $d_H(A_j) = 0$. Since A is a finite union of the $\{A_j\}_{j=1}^m$, $d_H(A) = 0$.

We let B be the union of sets B_j, where $b = (b_1, \ldots, b_m) \in B_j$ if $b_j = 1$ and the other components lie in C ($b_i \in C$ for $i \neq j$); the argument above shows that $d_H(B) = 0$, and so $K_m = A \cup B$ also has Hausdorff dimension zero.

Now let P be a projection (not necessarily orthogonal) of rank strictly less than m. Choose $v = (v_1, \ldots, v_m) \in \ker P$ with $|v_i| \leq 1$ for all i and $v_j = 1$ for some index $j \in \{1, \ldots, m\}$. We show that $v = b - a$ with $b \in B_j$ and $a \in A_j$: it will then follow that

$$0 = Pv = Pb - Pa \qquad \Rightarrow \qquad Pb = Pa,$$

and so P is not injective. We take for all $k \geq 0$

$$\begin{aligned} a_i^l = 0 \quad \text{and} \quad b_i^l &= v_i^l \qquad l \in (M_{2k}, M_{2k+1}], \\ a_i^l = (v_i^l + 1) \mod 2 \quad \text{and} \quad b_i^l &= 1 \qquad l \in (M_{2k+1}, M_{2k+2}]. \end{aligned}$$

Clearly $a \in A_j$, $b \in B_j$, and $v = b - a$.

Given an infinite-dimensional Hilbert space H, take a countable orthonormal set $\{e_j\}_{j=1}^\infty$, and let K'_m be the subset of H obtained from K_m by identifying the coordinate axes of $\mathbb{R}^m$ with $\{e_j\}_{j=1}^m$. Set

$$K = \{0\} \cup \bigcup_{m=1}^{\infty} 2^{-m} K'_m.$$

Then K is a compact subset of H with $d_H(K) = 0$.

Now suppose that L is a linear map such that $L : H \to \mathbb{R}^k$ is injective on K. This provides a linear mapping from $\mathbb{R}^m$ into $\mathbb{R}^k$ that is injective on K_m, and using the decomposition lemma (Lemma 6.1) this yields a rank k projection in $\mathbb{R}^m$ that is injective on K_m, a contradiction.

In the light of the result of Theorem 6.2, the set K has $d_H(K) = 0$ but $d_H(K - K) = \infty$.

6.2 Embedding sets with $d_H(X - X)$ finite

We now prove that linear embeddings do exist when $d_H(X - X)$ is finite. Mañé showed that under this condition a generic set of projections onto any subspace of $\mathscr{B}$ of dimension greater than $d_H(X - X)$ are one-to-one. The result here provides a version of his result in terms of prevalence, and replaces such projections by linear maps into $\mathbb{R}^k$. Given that we have already set up the machinery of prevalence and proved the inequality (5.8), the proof here is much simpler than Mañé's.

Theorem 6.2 *Let X be a compact subset of a real Banach space $\mathscr{B}$ such that $d_H(X - X) < k$, where k is a positive integer. Then a prevalent set of linear maps $L : \mathscr{B} \to \mathbb{R}^k$ are one-to-one between X and its image.*

We use the notation $\|\cdot\|_*$ to denote the norm in $\mathscr{B}^*$.

Proof Let V_n be a sequence of linear subspaces of $\mathscr{B}^*$ defined as follows. For each n, cover the set

$$Z_n = \{z \in X - X : \|z\| \geq 2^{-n}\}$$

using a collection of balls of radius $2^{-(n+1)}$ whose centres z_j lie in Z_n. Since Z_n is compact, there are a finite number of these balls.

Now, using the Hahn–Banach Theorem, there exists a corresponding set ψ_j of elements of $\mathscr{B}^*$ such that $\psi_j(z_j) = \|z_j\|$ and $\|\psi_j\|_* = 1$. Observe that for any $z \in Z_n$, there exists a j such that $z \in B(z_j, 2^{-(n+1)})$, and hence

$$|\psi_j(z)| = |\psi_j(z - z_j) + \psi_j(z_j)| \geq \|z_j\| - \|z - z_j\| \geq 2^{-(n+1)}.$$

Let V_n be the subspace of $\mathscr{B}^*$ spanned by the $\{\psi_j\}$, and write $d_n = \dim(V_n)$. Let $\mathscr{V} = \{V_n\}_{n=1}^{\infty}$ and for any $\gamma > 1$ let $E = E_\gamma(\mathscr{V})$ and let μ be the associated probability measure as defined in Section 5.2.

Now take $f \in \mathscr{L}(\mathscr{B}, \mathbb{R}^k)$, and let M be a Lipschitz constant valid for all $\{f + L : L \in E\}$. Let

$$Q_n = \{L \in E : (f + L)(z) = 0 \text{ for some } z \in Z_n\}$$

be the set of all linear maps in E for which $f + L$ fails to be injective for some pair $x, y \in X$ with $\|x - y\| \geq 2^{-n}$.

We will now show that $\sum_{n=1}^{\infty} \mu(Q_n) = 0$.

Choose $\delta > 0$, and for each n (which is taken to be fixed for this portion of the argument) cover Z_n with a collection of balls $B(z_j, \epsilon_j)$ such that

$$\sum_j \epsilon_j^k < 2^{-n}\delta \left(d_n^k n^{2k} 2^{nk}\right)^{-1}, \tag{6.1}$$

which is possible since $d_{\mathrm{H}}(Z_n) \leq d_{\mathrm{H}}(X - X) < k$ (see Lemma 2.7).

Let $Y_j = Z_n \cap B(z_j, \epsilon_j)$ and take $z_0 \in Y_j$. Then

$$|(f + L)(z_0)| > 2M\epsilon_j \quad \Rightarrow \quad |(f + L)(z)| > 0$$

for all $z \in Y_j$. The measure of

$$Q_{nj} = \{L \in E : (f + L)(z) = 0 \text{ for some } z \in Y_j\}$$

is therefore bounded by the measure of

$$\hat{Q}_{nj} = \{L \in E : |(f + L)(z_0)| \leq 2M\epsilon_j\}.$$

Now, since $z_0 \in Z_n$, by construction there exists a $\psi \in V_n$ with $\|\psi\|_* = 1$ such that $|\psi(z_0)| \geq 2^{-(n+1)}$, and Lemma 5.10 implies that

$$\mu(\hat{Q}_{nj}) \leq c(d_n n^2 2M\epsilon_j |\psi(z_0)|^{-1})^k,$$

whence

$$\mu(Q_{nj}) \leq c(4M)^k\, d_n^k n^{2k} 2^{nk} \epsilon_j^k.$$

Using (6.1) this implies that

$$\mu(Q_n) \leq \sum_j \mu(Q_{nj}) \leq c(4M)^k 2^{-n}\delta. \tag{6.2}$$

Now,

$$\bigcup_{n=1}^{\infty} Z_n = (X - X) \setminus \{0\},$$

and so

$$\bigcup_{n=1}^{\infty} Q_n = \{L \in E : (f + L)(z) = 0 \text{ for some nonzero } z \in X - X\}$$

is the set 'E_{bad}' of all $L \in E$ such that $f + L$ is not injective on X. It follows from (6.2) that

$$\mu(E_{bad}) \leq \sum_{n=1}^{\infty} \mu(Q_n) \leq c(4M)^k \delta.$$

Since $\delta > 0$ is arbitrary, $\mu(E_{bad}) = 0$ and the theorem is proved. □

6.3 No modulus of continuity is possible for L^{-1}

We now use a particular choice of orthogonal sequence in a Hilbert space (cf. Section 3.4) to show that $d_H(X - X) < \infty$ is not sufficient to guarantee any specified functional form of the modulus of continuity of L^{-1}. The argument is based on that of Ben-Artzi *et al.* (1993), who considered a similar question in the context of the upper box-counting dimension (see Section 8.2).

The following lemma is the key to this analysis (we will prove a more general version of this result later in Lemma 8.3).

Lemma 6.3 *Let P be any orthogonal projection in H, and $\{e_j\}_{j=1}^{\infty}$ any orthonormal subset of H. Then*

$$\text{rank } P \geq \sum_{j=1}^{\infty} \|Pe_j\|^2,$$

with equality guaranteed if $\{e_j\}_{j=1}^{\infty}$ is a basis for H.

Proof Suppose that P has rank k. Then there exists an orthonormal basis $\{u_1, \ldots, u_k\}$ for PH, so that for any $x \in H$,

$$Px = \sum_{j=1}^{k} (x, u_j) u_j.$$

In particular, $Pe_i = \sum_{j=1}^{k} (e_i, u_j) u_j$, so that

$$\|Pe_i\|^2 = (Pe_i, Pe_i) = (Pe_i, e_i) = \sum_{j=1}^{k} (e_i, u_j)(u_j, e_i) = \sum_{j=1}^{k} |(e_i, u_j)|^2.$$

It follows that

$$\sum_{i=1}^{\infty} \|Pe_i\|^2 = \sum_{i=1}^{\infty} \sum_{j=1}^{k} |(e_i, u_j)|^2 = \sum_{j=1}^{k} \sum_{i=1}^{\infty} |(e_i, u_j)|^2 \leq \sum_{j=1}^{k} \|u_j\|^2 = k,$$

with equality if the $\{e_i\}_{i=1}^{\infty}$ form a basis for H. □

Given any nondecreasing function $f : [0,\infty) \to [0,\infty)$ with $f(0) = 0$, we will show that there exists a compact set X with $d_{\rm H}(X - X) = 0$ such that the inequality

$$\|Pa\| \geq \epsilon f(\|a\|) \qquad \text{for all} \qquad a \in X \tag{6.3}$$

cannot hold for any $\epsilon > 0$ and any finite-rank orthogonal projection P.

The set X will be an orthogonal sequence of the form $\{\alpha_n e_n\}_{n=1}^{\infty} \cup \{0\}$, where $\{e_n\}$ is an orthonormal set in H. Note that $d_{\rm H}(X - X) = 0$, since $X - X$ is countable.

Suppose that (6.3) does hold. Then

$$\|P(\alpha_j e_j)\| = |\alpha_j|\|Pe_j\| \geq \epsilon f(|\alpha_j|) \qquad \text{for all} \qquad j = 1, \ldots,$$

i.e. $\|Pe_j\| \geq \epsilon f(\alpha_j)/\alpha_j$. Using Lemma 6.3 it follows that

$$\text{rank}(P) \geq \sum_{j=1}^{\infty} \|Pe_j\|^2 \geq \epsilon^2 \sum_{j=1}^{\infty} \left(\frac{f(\alpha_j)}{\alpha_j}\right)^2. \tag{6.4}$$

Now given any choice of f, set $\phi_n = nf(1/n)$, let N_n be the first integer greater than or equal to $1/\phi_n$, and define $T_j = \sum_{n=1}^{j} N_n$; for $T_j \leq i \leq T_{j+1}$ set $\alpha_i = 1/j$. This gives an orthogonal sequence X for which the right-hand side of (6.4) is infinite, and hence no finite-rank orthogonal projection can satisfy (6.3).

Since $0 \in X$, $X \subset X - X$; so there can be no finite-dimensional projection P for which

$$\|P(x_1 - x_2)\| \geq \epsilon f(\|x_1 - x_2\|) \qquad \text{for all} \qquad x_1, x_2 \in X,$$

for any value of $\epsilon > 0$. It follows from the decomposition lemma (Lemma 6.1) that if one can rule out such a modulus of continuity for orthogonal projections, the same follows for more general finite-rank linear maps.

Note that this argument also shows that one cannot prove a better embedding theorem than Theorem 6.2 if one strengthens the assumption to one on the modified box-counting dimension introduced in Exercise 3.6: all the above examples are countable sets, and so have modified box-counting dimension zero.

7
Thickness exponents

Theorem 4.3 gave an embedding result for subsets of $\mathbb{R}^N$ in terms of their upper box-counting dimension. As remarked at the end of Chapter 4, if we want to generalise the argument to subsets of infinite-dimensional spaces, we encounter a possible problem.

The proof of Theorem 4.3 relied on an application of the inequality

$$\mu\{L \in E : |\alpha + Lx| \leq \epsilon\} \leq cN^{k/2}\left(\frac{\epsilon}{|x|}\right)^k,$$

with $\epsilon = c2^{-n/\alpha}$ and $|x| \geq 2^{-n}$. In Chapter 6 we proved a generalised version of this inequality for subsets of a Hilbert space H,

$$\mu\{L \in E : |\alpha + Lx| \leq \epsilon\} \leq c\,(\dim V_j)^{k/2}\left(\frac{\epsilon}{\|P_j x\|}\right)^k, \tag{7.1}$$

where P_j is the orthogonal projection onto some subspace V_j of H used in the construction of E. If $\|x\| \geq 2^{-j}$ then we can ensure that $\|P_j x\|$ is bounded below by (a constant multiple of) 2^{-j} if we choose the space V_j appropriately. If

$$\text{dist}(X, V_j) \leq 2^{-j}/3$$

then, recalling that in the proof x was an element of the set of differences $X - X$, i.e. $x = x_1 - x_2$ with $x_1, x_2 \in X$, it follows that

$$\|P_j x\| = \|P_j(x_1 - x_2)\| \geq \|x_1 - x_2\| - \|x_1 - P_j x_1\| - \|x_2 - P_j x_2\| \geq 2^{-j}/3.$$

While this gives a lower bound on $\|P_j x\|$ of the required form, the dimension of V_j occurs in the estimate (7.1). In order to carry the argument through successfully, we will need some control on how the dimension of V_j grows with j. This is provided by the thickness exponent, $\tau(X)$, introduced by Hunt

& Kaloshin (1999), and discussed in Section 7.1. This exponent can be shown to be zero when the set X consists of C^∞ functions, see Lemma 13.1.

A related quantity which can be defined for subsets of Hilbert spaces is the Lipschitz deviation dev(X), covered in Section 7.2. Introduced by Olson & Robinson (2010) and refined further by Pinto de Moura & Robinson (2010b), this can replace the thickness exponent in the generalised (infinite-dimensional) version of Theorem 4.3, and can be shown to be zero for the attractors arising in the infinite-dimensional dynamical systems generated by a number of canonical partial differential equations (Section 13.2).

Finally, in Section 7.3 we define a version of the thickness, the 'dual thickness' $\tau^*(X)$, appropriate for subsets of Banach spaces. In a Hilbert space $\tau^*(X) \leq \text{dev}(X) \leq \tau(X)$; it is not clear how the thickness and dual thickness are related for subsets of Banach spaces, but one can prove the useful result that $\tau(X) = 0$ implies that $\tau^*(X) = 0$ (Proposition 7.10).

7.1 The thickness exponent

The 'thickness exponent' (or simply 'thickness') was introduced by Hunt & Kaloshin (1999), although a similar idea was used in the paper by Foias & Olson (1996) without leading to any formal definition.

Definition 7.1 Let X be a subset of a Banach space $\mathscr{B}$. The thickness exponent of X in $\mathscr{B}$, $\tau(X; \mathscr{B})$ is given by

$$\tau(X; \mathscr{B}) = \limsup_{\epsilon \to 0} \frac{\log d_{\mathscr{B}}(X, \epsilon)}{-\log \epsilon},$$

where $d_{\mathscr{B}}(X, \epsilon)$ is the dimension of the smallest linear subspace V of $\mathscr{B}$ such that

$$\text{dist}_{\mathscr{B}}(X, V) \leq \epsilon,$$

i.e. every point in X lies within ϵ of V (in the norm of $\mathscr{B}$).

We will usually drop the space $\mathscr{B}$ from the notation in what follows, preferring the simpler $d(X, \epsilon)$ and $\tau(X)$. But note that, as with the box-counting dimension, the definition depends on the space in which we consider X. We note here for use later that if $\mathscr{B}_1$ and $\mathscr{B}_2$ are two Banach spaces with $X \subset \mathscr{B}_1 \subseteq \mathscr{B}_2$, then

$$\|u\|_{\mathscr{B}_2} \leq c\|u\|_{\mathscr{B}_1} \quad \Rightarrow \quad \tau(X; \mathscr{B}_2) \leq \tau(X; \mathscr{B}_1). \tag{7.2}$$

As observed by Hunt & Kaloshin, the thickness is always bounded by the box-counting dimension.

Lemma 7.2 *If X is a subset of a Banach space $\mathscr{B}$ then $\tau(X) \le d_{\rm B}(X)$.*

Proof Given $\epsilon > 0$, cover X with $N(X, \epsilon)$ balls of radius ϵ. Then every point of X lies within ϵ of the linear subspace V that is spanned by the centres of these balls. (This is essentially the way that the idea was used by Foias & Olson (1996).) Since the dimension of V is no greater than $N(X, \epsilon)$, this implies that $d(X, \epsilon) \le N(X, \epsilon)$ and the lemma follows. □

We now show that for the example of an orthogonal sequence in a Hilbert space (as considered in Lemma 3.5), the thickness is in fact equal to the box-counting dimension. To show this we will require the following lemma due to M. Doré (personal communication).

Lemma 7.3 *Let $X = \{v_1, \ldots, v_n\}$ be an orthogonal set in a Hilbert space H. Then*

$$d(X, \epsilon) \ge n(1 - \epsilon^2/M^2),$$

where $M = \min\{\|v_1\|, \ldots, \|v_n\|)$.

Proof If $d(X, \epsilon) = d$ then there exist $v_i' \in H$ such that $\|v_i' - v_i\| < \epsilon$, and such that the space spanned by $\{v_1', \ldots, v_n'\}$ has dimension d. Let P be the orthogonal projection onto U, the n-dimensional space spanned by $\{v_1, \ldots, v_n\}$ and let $v_i'' = Pv_i'$. Since $Pv_i = v_i$ we still have the inequality $\|v_i'' - v_i\| < \epsilon$ and clearly the dimension of the linear span of $\{v_1', \ldots, v_n'\}$ is at least that of the linear span of $\{v_1'', \ldots, v_n''\}$.

Suppose that the linear span of $\{v_1'', \ldots, v_n''\}$ has dimension $n - r$. We can write any element of U in terms of the $\{v_j''\}$ and an orthonormal basis for their r-dimensional orthogonal complement in U, $\{u_1, \ldots, u_r\}$. So

$$\begin{aligned} n\epsilon^2 &\ge \sum_{i=1}^{n} \|v_i'' - v_i\|^2 \ge \sum_{i=1}^{n}\sum_{j=1}^{r} |(v_i, u_j)|^2 \\ &= \sum_{j=1}^{r}\sum_{i=1}^{n} \|v_i\|^2 \left|(u_j, \frac{v_i}{\|v_i\|})\right|^2 \\ &\ge M^2 \sum_{j=1}^{r}\sum_{i=1}^{n} \left|(u_j, \frac{v_i}{\|v_i\|})\right|^2 = M^2 r. \end{aligned}$$

It follows that $d(X, \epsilon) \ge n(1 - \epsilon^2/M^2)$ as claimed. □

We now use this to find an expression for $\tau(A)$ when A is an orthogonal sequence (the proof follows Pinto de Moura & Robinson (2010a)).

Lemma 7.4 *Let $A = \{a_n e_n\}_{n=1}^{\infty} \cup \{0\}$, where $\{e_j\}_{j=1}^{\infty}$ is an orthonormal subset of a Hilbert space H, and $a_n \to 0$ with $|a_{n+1}| \le |a_n|$. Then*

$$\tau(A) = \limsup_{n\to\infty} \frac{\log n}{-\log |a_n|}. \tag{7.3}$$

Proof Combining the results of Lemmas 3.5 and 7.2 shows that $\tau(A)$ is bounded by the right-hand side of (7.3).

The argument leading to the reverse inequality is similar to that used for Lemma 3.5. Choose n large enough that $|a_n| < 1$, denote by n' the unique integer $n' \ge n$ such that

$$|a_n| = |a_{n+1}| = \cdots = |a_{n'}| > |a_{n'+1}|,$$

and set $\epsilon_n^2 = (|a_{n'}|^2 + |a_{n'+1}|^2)/4$. Since $|a_{n'}|^2 > 2\epsilon_n^2$

$$1 - \frac{\epsilon_n^2}{|a_{n'}|^2} > \tfrac{1}{2},$$

and so Lemma 7.3 implies that

$$d(A, \epsilon_n) \ge n' \left(1 - \frac{\epsilon_n^2}{|a_{n'}|^2}\right) > \frac{n'}{2}.$$

Combining this inequality with $2\epsilon_n > |a_{n'}|$, $n' \ge n$, and $|a_n| = |a_{n'}|$, we obtain

$$\tau(A) \ge \limsup_{n\to\infty} \frac{\log d(A, \epsilon_n)}{-\log \epsilon_n} \ge \limsup_{n\to\infty} \frac{\log(n/2)}{\log(2/|a_n|)} \ge \limsup_{n\to\infty} \frac{\log n}{-\log |a_n|}. \quad \square$$

Friz & Robinson (1999) showed that if U is a sufficiently regular bounded domain in $\mathbb{R}^n$ and X is a subset of $L^2(U)$ that consists of functions that are uniformly bounded in the Sobolev space $H^s(U)$, it follows that $\tau(X) \le n/s$, see Lemma 13.1. In particular this shows that the attractors of partial differential equations that are 'smooth' (bounded in H^s for all s) have thickness exponent zero.

7.2 Lipschitz deviation

The m-Lipschitz deviation was introduced by Olson & Robinson (2010) as a first step towards generalising the thickness exponent. Denote by $\delta_m(X, \epsilon)$ the smallest dimension of a linear subspace U of H such that

$$\mathrm{dist}(X, G_U[\phi]) < \epsilon$$

for some m-Lipschitz function $\phi : U \to U^{\perp}$,

$$\|\phi(u) - \phi(v)\| \leq m\|u - v\| \qquad \text{for all} \qquad u, v \in U,$$

where $U^{\perp}$ is the orthogonal complement of U in H and $G_U[\phi]$ is the graph of ϕ over U:

$$G_U[\phi] = \{ u + \phi(u) : u \in U \}.$$

The m-Lipschitz deviation is given by

$$\mathrm{dev}_m(X) = \limsup_{\epsilon \to 0} \frac{\log \delta_m(X, \epsilon)}{-\log \epsilon}.$$

The Lipschitz deviation of X (Pinto de Moura & Robinson, 2010b) is

$$\mathrm{dev}(X) = \lim_{m \to \infty} \mathrm{dev}_m(X);$$

since $\mathrm{dev}_m(X)$ is nonincreasing in m the limit clearly exists provided that $\mathrm{dev}_m(X)$ is finite for some $m > 0$.

Note that dev(X) is bounded above by $\tau(X)$, since $\mathrm{dev}_m(X) \leq \tau(X)$ for every $m > 0$ (one can always approximate by the graph of the zero function, which is m-Lipschitz). We now show, following Pinto de Moura & Robinson (2010b), that this inequality can be strict.

7.2.1 An example with dev(X) < $\tau(X)$.

Let $\{e_j\}_{j=1}^{\infty}$ be an orthonormal set in a Hilbert space H, and consider the set

$$X = \left\{ \frac{1}{n} e_1 + \frac{1}{n^2} e_n : n \geq 2 \right\} \cup \{0\}.$$

It is relatively easy to show that X is contained in the graph of a 3-Lipschitz function of the one-dimensional subspace E_1 spanned by e_1: define ϕ on the discrete set of points $\{e_1/n\}_{n \in \mathbb{N}} \cup \{0\}$ by

$$\phi(e_1/n) = \frac{e_n}{n^2} \qquad n \geq 2 \qquad \text{and} \qquad \phi(0) = 0.$$

On its domain of definition, ϕ is Lipschitz: for $m > n$,

$$\begin{aligned} |\phi(e_1/n) - \phi(e_1/m)| &= \left| \frac{e_n}{n^2} - \frac{e_m}{m^2} \right| \\ &= n^{-2} + m^{-2} \leq n^{-2} + (n+1)^{-2} < \frac{3}{n(n+1)} \end{aligned}$$

and

$$\left| \frac{e_1}{n} - \frac{e_1}{m} \right| = \left| \frac{1}{n} - \frac{1}{m} \right| > \left| \frac{1}{n} - \frac{1}{n+1} \right| = \frac{1}{n(n+1)},$$

and so

$$|\phi(e_1/n) - \phi(e_1/m)| \leq 3\left|\frac{e_1}{n} - \frac{e_1}{m}\right|.$$

The function ϕ can be extended to a 3-Lipschitz function defined on the whole of E_1 (see Wells & Williams (1975), for example). It follows that $\text{dev}_3(X) = 0$, and so $\text{dev}(X) = 0$.

We now show that $\tau(X) \geq 1$, following the argument used above to prove Lemma 7.4. For $n \geq 1$ set

$$a_n = \frac{e_1}{n+1} + \frac{e_{n+1}}{(n+1)^2};$$

note that $\|a_n\| \geq \|a_{n+1}\|$ and $\lim_{n\to\infty} \|a_n\| = 0$. Let $X = \{a_1, a_2, \ldots\}$. Set $\epsilon_n^2 = (\|a_n\|^2 + \|a_{n+1}\|^2)/4$. Since $\|a_j\|^2 \geq \|a_n\|^2 > 2\epsilon_n^2$ for $j = 1, \ldots, n$, it follows from the above lemma that

$$d(X, \epsilon_n) \geq d(\{a_1, \ldots, a_n\}, \epsilon_n) \geq n\left(1 - \frac{\epsilon_n^2}{\|a_n\|^2}\right) \geq \frac{n}{2}.$$

Since $(n+1)^{-1} < \|a_n\| < 2\epsilon_n$,

$$\tau(X) \geq \limsup_{n\to\infty} \frac{\log d(X, \epsilon_n)}{-\log \epsilon_n} \geq \limsup_{n\to\infty} \frac{\log(n/2)}{\log 2(n+1)} = 1.$$

7.3 Dual thickness

The definition of the Lipschitz deviation requires a splitting of the space into a finite-dimensional subspace U and its orthogonal complement. In a Banach space it is not obvious how to perform such a splitting. Instead we define yet another new quantity, the 'dual thickness'. We will see that in a Hilbert space this is bounded by the Lipschitz deviation, so in this setting offers a further refinement of the thickness exponent.

The definition is based on the construction used in the proof of Theorem 6.2, and encodes precisely the property of 'approximation' that is needed in the argument used to prove the embedding theorem that follows in the next chapter (Theorem 8.1).

Definition 7.5 Given $\theta > 0$, let $n_\theta(X, \epsilon)$ denote the lowest dimension of any linear subspace V of $\mathscr{B}^*$ such that for any $x, y \in X$ with $\|x - y\| \geq \epsilon$ there exists an element $\psi \in V$ such that $\|\psi\|_* = 1$ and

$$|\psi(x - y)| \geq \epsilon^{1+\theta}.$$

Set

$$\tau_\theta^*(X) = \limsup_{\epsilon\to 0} \frac{\log n_\theta(X,\epsilon)}{-\log \epsilon},$$

and define the *dual thickness* $\tau^*(X)$ by

$$\tau^*(X) = \lim_{\theta\to 0} \tau_\theta(X).$$

It is also useful to introduce the following more straightforward definition.

Definition 7.6 Given $\theta > 0$, let $m_\alpha(X,\epsilon)$ denote the lowest dimension of any linear subspace V of $\mathscr{B}^*$ such that for any $x, y \in X$ with $\|x - y\| \geq \epsilon$ there exists an element $\psi \in V$ such that $\|\psi\|_* = 1$ and

$$|\psi(x - y)| \geq \alpha\epsilon.$$

Define

$$\sigma_\alpha^*(X) = \limsup_{\epsilon\to 0} \frac{\log m_\alpha(X,\epsilon)}{-\log \epsilon}.$$

It would now be natural to define $\sigma^*(X) = \lim_{\alpha\to 0} \sigma_\alpha^*(X)$; this gives another possible definition of a 'dual thickness', but with the more tortuous definition of τ^* (which is never larger than σ^*, see below) we can still prove an embedding theorem, and more importantly we can show that zero thickness (in the sense of Hunt & Kaloshin's definition) implies zero dual thickness (Proposition 7.10); this does not seem to be possible using σ^*.

The following lemma shows that $\sigma_\alpha^*(X)$ provides an upper bound for $\tau^*(X)$ for any $\alpha > 0$.

Lemma 7.7 *If X is a compact subset of a Banach space $\mathscr{B}$ then $\tau^*(X) \leq \sigma_\alpha^*(X)$ for any $\alpha > 0$.*

Proof If V is an n-dimensional subspace of $\mathscr{B}^*$ such that for all $x, y \in X$ with $\|x - y\| \geq \epsilon$ there exists a $\psi \in V$ with $\|\psi\|_* = 1$ and $|\psi(x - y)| \geq \alpha\epsilon$, then it is clear that

$$|\psi(x - y)| \geq \epsilon^{1+\theta}$$

for all ϵ small enough that $\epsilon^\theta < \alpha$, and so $\tau_\theta^*(X) \leq \sigma_\alpha^*(X)$ for all $\theta > 0$. □

The following simple corollary shows that $\tau^*(X) \leq d_B(X)$, i.e. that the dual thickness is well adapted for use with sets that have finite box-counting dimension. Corollary 8.4 shows that there are sets for which this upper bound is attained, so $\tau^*(X)$ is not always zero (there are possible definitions of 'thickness exponents' that one might expect to be useful but which turn out to be zero whenever X has finite box-counting dimension, see Exercise 7.2 for one example).

Corollary 7.8 *Let X be a compact subset of a Banach space $\mathscr{B}$; then $\tau^*(X) \leq d_B(X)$.*

Proof Take $d > d_B(X)$. Then there exists an $\epsilon_0 > 0$ such that for all $\epsilon < \epsilon_0$, X can be covered with a collection $B(x_j, \epsilon/12)$ of $N \leq \epsilon^{-d}$ balls, where the x_j are chosen to be linearly independent. To see that this is possible, first cover X with a collection of balls $B(z_j, \epsilon/13)$. This is a finite collection; since $\mathscr{B}$ is infinite-dimensional one can perturb each z_j in turn to some x_j such that the resulting collection $\{x_1, \ldots, x_n\}$ is linearly independent for each $n \leq N$. One can then enlarge slightly the radius of each ball.

Now use the Hahn–Banach Theorem to define a collection of linear functionals ψ_j with the property

$$\psi_j(x_i) = \delta_{ij}\|x_j\| \qquad \text{and} \qquad \|\psi_j\|_* = 1,$$

and let V be the subspace of $\mathscr{B}^*$ spanned by the ψ_j.

Now, given $x, y \in X$ with $\|x - y\| \geq \epsilon$, there exist x_j, x_k such that

$$\|x - x_j\| \leq \frac{\epsilon}{12} \qquad \text{and} \qquad \|y - x_k\| \leq \frac{\epsilon}{12},$$

and so in particular $\|x_j - x_k\| \geq 5\epsilon/6$.

Clearly $\psi_j - \psi_k \in V$, and

$$1 \leq \|\psi_j - \psi_k\|_* \leq 2,$$

so that $(\psi_j - \psi_k)/\|\psi_j - \psi_k\|_*$ is an element of V with norm 1. We have

$$\begin{aligned}
\frac{\psi_j - \psi_k}{\|\psi_j - \psi_k\|_*}(x - y) &= \frac{\psi_j - \psi_k}{\|\psi_j - \psi_k\|_*}[(x - x_j) + x_j - x_k + (x_k - y)] \\
&\geq -\frac{\epsilon}{12} + \frac{\|x_j\| + \|x_k\|}{\|\psi_j - \psi_k\|_*} - \frac{\epsilon}{12} \\
&\geq \frac{\|x_j - x_k\|}{2} - \frac{\epsilon}{6} \\
&\geq \frac{5\epsilon}{12} - \frac{\epsilon}{6} = \frac{\epsilon}{4}.
\end{aligned}$$

It follows that $\sigma^*_{1/4}(X) \leq d$, and since $d > d_B(X)$ was arbitrary, $\tau^*(X) \leq \sigma^*_{1/4}(X) \leq d_B(X)$. □

In a Hilbert space we can do better than this, showing that the dual thickness is bounded by the Lipschitz deviation (and hence by the thickness).

Lemma 7.9 *If X is a compact subset of a Hilbert space H then $\tau^*(X) \leq \mathrm{dev}(X) \leq \tau(X)$.*

Proof In fact the argument here shows that $\sigma^*_{1/6m}(X) \leq \mathrm{dev}_m(X)$, and the result as stated is a consequence of Lemma 7.7.

Suppose that $X \subset H$ and V is a linear subspace of H such that there exists an m-Lipschitz function $\phi : V \to V^\perp$ with

$$\mathrm{dist}(X, G_V[\phi]) \leq \epsilon/6m,$$

where $m \geq 1$. It follows that for each $x \in X$ there exists a $p \in V$ such that $\|x - (p + \phi(p))\| \leq \epsilon/6m$. Writing P for the orthogonal projection onto V, and $Q = I - P$,

$$\begin{aligned}\|x - (Px + \phi(Px))\| &= \|Qx - \phi(Px)\| \\ &\leq \|Qx - \phi(p)\| + \|\phi(p) - \phi(Px)\| \\ &\leq \|Qx - \phi(p)\| + m\|p - Px\| \\ &\leq 2m\|x - (p + \phi(p))\| \leq \epsilon/3.\end{aligned}$$

Now, for any $u, v \in H$,

$$\|(Pu + \phi(Pu)) - (Pv + \phi(Pv))\| \leq 2m\|Pu - Pv\|$$

and if $u, v \in X$ with $\|u - v\| \geq \epsilon$, then

$$\begin{aligned}&\|(Pu + \phi(Pu)) - (Pv + \phi(Pv))\| \\ &\quad \geq \|u - v\| - \|Qu - \phi(Pu)\| - \|Qv - \phi(Pv)\| \\ &\quad \geq \epsilon/3,\end{aligned}$$

which implies that $\|Pu - Pv\| \geq \epsilon/6m$.

The subspace V has a natural isometric linear embedding into H^* via the mapping $u \mapsto \varphi_u$, where

$$\varphi_u(x) = (x, u).$$

Given $u, v \in X$ with $\|u - v\| \geq \epsilon$, let $d = P(u - v)/\|P(u - v)\|$ and $\psi = \varphi_d$. Clearly $\|\psi\|_* = 1$ and $|\psi(u - v)| \geq \epsilon/6m$. □

In a Banach space the relationship between the thickness and dual thickness is not clear in general. Nevertheless, it is possible to show that sets that have 'zero thickness' according to Hunt & Kaloshin's definition also have zero dual thickness (this is the reason for the slightly torturous definition of the dual thickness). This will prove particularly useful in Chapter 15.

Proposition 7.10 *Let X be a compact subset of a Banach space $\mathscr{B}$. If $\tau(X) = 0$ then $\tau^*(X) = 0$.*

Proof Take any $\tau \in (0, 1)$. It follows from the definition of the thickness exponent that for any $\beta > 0$, for every ϵ with $0 < \epsilon < 1$ there exists a subspace U of dimension $n \leq c_\tau \epsilon^{-\beta\tau}$ such that $\text{dist}(X, U) \leq \epsilon^\beta$.

Let P be a projection onto U with $\|P\| \leq n$; the existence of such a projection is guaranteed by Exercise 7.4. Given any $x \in \mathscr{B}$, if $\text{dist}(x, U) = \delta$ there exist $u \in U$ and $y \in \mathscr{B}$ with $\|y\| = \delta$ such that $x = u + y$. It follows, since the norm of $Q = I - P$ is bounded by $1 + n$, that

$$\begin{aligned}\|x - Px\| &= \|Qx\| = \|Q(u + y)\| = \|Qy\| \leq \|Q\|\|y\| \leq (1 + n)\text{dist}(x, U) \\ &\leq 2c_\tau\epsilon^{-\beta\tau}\epsilon^\beta = c'_\tau\epsilon^{\beta(1-\tau)}.\end{aligned}$$

Choosing $\beta = 1/(1 - \tau)$ it follows that for any $\epsilon > 0$ there exists a space U of dimension $n \leq c''_\tau\epsilon^{-\tau/(1-\tau)}$ and a projection P onto U with $\|P\| \leq n$ such that

$$\sup_{x \in X} \|x - Px\| \leq \epsilon/3.$$

Now let V be the n-dimensional linear subspace of $\mathscr{B}^*$ given by

$$V = \{L \circ P : L \in U^*\},$$

where, as above, U^* is the dual of U, which is also n-dimensional. Given any $x, y \in X$ with $\|x - y\| \geq \epsilon$, one can find an $L \in U^*$ such that $\|L\|_* = 1$ and $L(P(x - y)) = \|P(x - y)\| \geq \epsilon/3$. It follows that there exists a $\psi \in V$, namely $\psi = L \circ P$, such that

$$\|\psi\|_* \leq n \qquad \text{and} \qquad |\psi(x - y)| \geq \epsilon/3.$$

Rescaling ψ by a factor of $\|\psi\|_* \leq n \leq c''\epsilon^{-\tau/(1-\tau)}$, it follows that there exists a $\psi \in V$ with $\|\psi\|_* = 1$ such that

$$|\psi(x - y)| \geq c\,\epsilon^{1+(\tau/(1-\tau))}.$$

Thus

$$\tau^*_{\tau/(1-\tau)}(X) \leq \frac{\tau}{1 - \tau},$$

and since $\tau/(1 - \tau) \to 0$ as $\tau \to 0$, $\tau^*(X) = 0$ as claimed. □

Exercises

7.1 Let X be a subset of a Banach space $\mathscr{B}$, and denote by $\varepsilon(X, n)$ the minimum distance between X and any n-dimensional linear subspace of $\mathscr{B}$. Show that

$$\tau(X) \le \limsup_{n\to\infty} \frac{\log n}{-\log \varepsilon(X,n)}. \tag{7.4}$$

(One can in fact prove equality here, see Lemma 2 in Kukavica & Robinson (2004).)

7.2 One could try to define another 'thickness measure' for a set $X \subset H$ as follows. For each $\epsilon > 0$, let $d_{\mathrm{LE}}(X,\epsilon)$ be the smallest n such that there exists a 2-Lipschitz map $\phi : \mathbb{R}^n \to H$ such that $\mathrm{dist}(X, \phi(\mathbb{R}^n)) < \epsilon$. Define

$$\tau_{\mathrm{LE}}(X) = \limsup_{\epsilon\to 0} \frac{\log d_{\mathrm{bL}}(X,\epsilon)}{-\log \epsilon}.$$

The Johnson–Lindenstrauss Lemma (Johnson & Lindenstrauss, 1984) guarantees that given a set of m points in a Hilbert space H, and an $n > O(\ln m)$, there is a function $f : H \to \mathbb{R}^n$ such that

$$\tfrac{1}{2}\|u - v\| \le |f(u) - f(v)| \le 2\|u - v\|.$$

Show that $\tau_{\mathrm{LE}}(X) = 0$ for any set X with $d_{\mathrm{B}}(X)$ finite.

7.3 Show that if U is a finite-dimensional Banach space $\dim(U) = n$ then there exists an 'Auerbach basis' for U: a basis $\{e_1, \ldots, e_n\}$ for U and corresponding elements $\{f_1, \ldots, f_n\}$ of U^* such that $\|e_j\| = \|f_j\|_* = 1$ for $j = 1, \ldots, n$ and $f_i(e_j) = \delta_{ij}, i, j = 1 \ldots, n$. [Hint: by identifying U with $(\mathbb{R}^n, \|\cdot\|)$ one can work in $\mathbb{R}^n$. For $x_1, \ldots, x_n \in \mathbb{R}^n$ let $\det(x_1, \ldots, x_n)$ denote the determinant of the $n \times n$ matrix with columns formed by the vectors $\{x_j\}$. Choose $\{e_1, \ldots, e_n\}$ with $\|e_j\| = 1$ such that $\det(e_1, \ldots, e_n)$ is maximal. Define candidates for the $\{f_j\}$ and check that the f_j satisfy the required properties.]

7.4 Use the result of the previous exercise to show that if U is any n-dimensional subspace of a Banach space $\mathscr{B}$, there exists a projection P onto U whose norm is no larger than n, $\|P\| \le n$.

8
Embedding sets of finite box-counting dimension

This chapter provides a proof of an infinite-dimensional version of Theorem 4.3: we prove the existence of linear embeddings into $\mathbb{R}^k$ for subsets of infinite-dimensional (Hilbert or Banach) spaces with finite upper box-counting dimension, and show that these linear maps have Hölder continuous inverses. This degree of smoothness allows for interesting corollaries for the attractors of infinite-dimensional dynamical systems, as discussed in Part II.

8.1 Embedding sets with Hölder continuous parametrisation

We prove a result that makes use of the dual thickness $\tau^*(X)$; note that if the dual thickness is zero then the Hölder exponent in (8.1) can be made arbitrarily close to 1 by choosing an embedding space of sufficiently high dimension. In light of this it is worth recalling from the previous chapter that

$$\tau^*(X) \leq \text{dev}(X) \leq \tau(X)$$

in a Hilbert space (Lemma 7.9), and that in a Banach space

$$\tau(X) = 0 \quad \Rightarrow \quad \tau^*(X) = 0$$

(Proposition 7.10).

Theorem 8.1 *Let X be a compact subset of a real Banach space $\mathscr{B}$, with $d_{\mathrm{B}}(X) = d < \infty$ and $\tau^*(X) = \tau$. Then for any integer $k > 2d$ and any θ with*

$$0 < \theta < \frac{k - 2d}{k(1 + \alpha\tau)}, \tag{8.1}$$

where $\alpha = 1/2$ if $\mathscr{B}$ is a Hilbert space and $\alpha = 1$ if $\mathscr{B}$ is a general Banach space, there exists a prevalent set of bounded linear maps $L : \mathscr{B} \to \mathbb{R}^k$ such *that*

$$\|x - y\| \leq C_L |Lx - Ly|^\theta \qquad \textit{for all} \qquad x, y \in X. \tag{8.2}$$

In particular, L is injective on X.

Foias & Olson (1996) first proved a result along these lines. They showed that in a Hilbert space there is a dense set of orthogonal projections that are injective on X and have a Hölder inverse, but did not give any explicit bound on the Hölder exponent. The proof of the theorem given here, almost identical to that of Theorem 4.3, is due essentially to Hunt & Kaloshin (1999), but incorporates the generalised estimate of Lemma 5.10 and the dual thickness, following Robinson (2009). Separating the more geometric elements of the proof (the estimates contained in Section 5.2) serves to clarify the argument.

Proof If θ satisfies (8.1), then there exist $\beta > 0$, $\sigma > \tau_\beta^*(X)$, and $\delta > d$ such that

$$0 < \theta < \frac{k - 2\delta}{k(1 + \beta + \alpha\sigma)}. \tag{8.3}$$

Since $\sigma > \tau_\beta^*(X)$, there exists a subspace of $\mathscr{B}^*$, V_j, of dimension $d_j \leq C_1 2^{j\theta\sigma}$ such that for any $x, y \in X$ with $\|x - y\| \geq 2^{-j\theta}$, one can find a $\psi \in V_j$ with

$$\|\psi\|_* = 1 \qquad \text{and} \qquad |\psi(x - y)| \geq 2^{-j\theta(1+\beta)}. \tag{8.4}$$

With $\mathscr{V} = \{V_j\}_{j=1}^\infty$, choose $\gamma > 1$ and let $E = E_\gamma(\mathscr{V})$ and μ the corresponding probability measure as defined in Section 5.2.

Now let

$$Z_j = \{z \in X - X : \|z\| \geq 2^{-\theta j}\},$$

and for a fixed choice of $f \in \mathscr{L}(\mathscr{B}, \mathbb{R}^k)$ let Q_j be the set of all those linear maps in E for which (8.2), with L replaced by $f + L$, fails for some $z \in Z_j$,

$$Q_j = \{L \in E : |(f + L)(z)| \leq 2^{-j} \text{ for some } z \in Z_j\}.$$

Since $d_B(X - X) \leq 2d_B(X) < 2\delta$, $X - X$ can be covered with no more than $C_2 2^{2j\delta}$ balls of radius 2^{-j}. Let Y be the intersection of Q_j with one of these balls.

Let M be a Lipschitz constant that is valid for all $f + L$, $L \in E$. If $z, z_0 \in Y$, then since $\|z - z_0\| \leq 2^{-(j-1)}$,

$$|(f + L)(z_0)| > (2M + 1)2^{-j} \quad \Rightarrow \quad |(f + L)(z)| > 2^{-j} \quad \text{for all } z \in Y.$$

Thus $\mu(Y)$ is bounded by the μ-measure of those $L \in E$ for which

$$|(f + L)(z_0)| \leq (2M + 1)2^{-j}.$$

It follows from Lemma 5.10 (in the Banach space case) or Lemma 5.6 (in the Hilbert space case) that

$$\mu(Y) \le [(2M+1)2^{-j} j^2 C_1^\alpha 2^{j\theta\sigma\alpha} |g(z_0)|^{-1}]^k$$

for any $g \in S_j$ (recall that $d_j \le C_1 2^{j\theta\sigma}$). Using the definition of the dual thickness there exists a $\psi \in S_j$ such that

$$|\psi(z_0)| \ge \|z_0\|^{1+\beta} \ge 2^{-\theta(1+\beta)j}$$

(cf. (8.4)), and so

$$\mu(Y) \le [(2M+1)2^{-j} j^2 C_1^\alpha 2^{j\theta\sigma\alpha} 2^{\theta(1+\beta)j}]^k.$$

Since Q_j is covered by no more than $C_2 2^{2j\delta}$ balls, we obtain

$$\begin{aligned}\mu(Q_j) &\le C_2 2^{2j\delta}[(2M+1)2^{-j} j^2 C_1^\alpha 2^{j\theta\sigma\alpha} 2^{\theta(1+\beta)j}]^k \\ &= C_3 j^{2k} 2^{-j[k(1-\theta(1+\beta+\sigma\alpha))-2\delta]}.\end{aligned}$$

The assumption (8.3) implies that

$$k(1-\theta(1+\beta+\sigma\alpha)) - 2\delta > 0,$$

and so the sum $\sum_{j=1}^\infty \mu(Q_j)$ is finite. It follows from the Borel–Cantelli Lemma (Lemma 4.2) that μ-almost every L belongs to only finitely many of the Q_j, which implies (8.2) for some appropriate constant C_L (the argument is identical to the concluding part of the proof of Theorem 4.3). □

One could try to repeat this proof, replacing the upper box-counting dimension by the lower box-counting dimension, but the assumption that $d_{\rm LB}(X) < \infty$ is not enough to show that L^{-1} is Hölder, as we will now see.

8.2 Sharpness of the Hölder exponent

As the dimension of the embedding space (k) in the above theorem tends to infinity, one obtains in (8.1) a limiting Hölder exponent $1/(1+\tau^*)$ in the Banach space case, or $1/(1+(\tau^*/2))$ in the Hilbert space case.

Hunt & Kaloshin (1999) provide an example showing that this limiting Hölder exponent is sharp in terms of the thickness: an involved construction based on taking particular paths through a binary tree yields a subset of ℓ^p for any $1 \le p < \infty$, such that $d_{\rm H}(X) = d_{\rm B}(X) = d$ and with the property that if $L : \ell^p \to \mathbb{R}^N$ is any bounded linear map then

$$d_{\rm H}(L(X)) \le \frac{d}{1+d/q}$$

with q the conjugate exponent of p. It follows from the behaviour of the Hausdorff dimension under θ-Hölder maps (Proposition 2.8(iv)) that one must therefore have $\theta < 1/(1 + d/q)$, cf. (8.8).

It is interesting that the 'sharpness' of the bound on the Hölder exponent must always be understood in terms of the quantities that are being used in the bound: for example, while Hunt & Kaloshin used their example to show that in a Hilbert space the bound

$$0 < \theta < \frac{k - 2d}{k(1 + \tau(X)/2)}$$

is sharp in terms of the thickness, we know from Theorem 8.1 that this bound can be improved in that one can substitute a different quantity for the thickness, e.g. the Lipschitz deviation or the dual thickness.

Here we show that the 'orthogonal sequence' considered in Lemma 3.5 provides a much simpler example that proves the sharpness of the Hölder exponent (in terms of the dual thickness). In order to do this we require Banach-space versions of Lemma 6.1 (the decomposition lemma) and Lemma 6.3 (relating the rank of a projection P to the images of an orthonormal set under P).

The following result can be found in Roman (2007, Theorem 3.5).

Lemma 8.2 *Let $\mathscr{B}$ be a Banach space. Suppose that $L : \mathscr{B} \to \mathbb{R}^k$ is a surjective linear map with $L(\mathscr{B}) = \mathbb{R}^k$ and V is the kernel of L. Then the quotient space $U = \mathscr{B}/V$ has dimension k, and L can be decomposed uniquely as MP, where P is a projection onto U and $M : U \to \mathbb{R}^k$ is an invertible linear map.*

The proof of the following version of Lemma 6.3, applicable to projections in ℓ^p, is due to Pinto de Moura (see Pinto de Moura & Robinson (2010a)).

Lemma 8.3 *Let P be a finite rank projection in ℓ^p ($1 < p < \infty$) or c_0 (in which case we take $p = \infty$). Then*

$$\text{rank } P \geq \left\{ \sum_{j=1}^{\infty} \|Pe_j\|_{\ell^p}^q \right\}^{1/q},$$

where $\{e_j\}_{j=1}^{\infty}$ is the canonical basis of ℓ^p (or c_0), and q is the conjugate exponent to p.

Note that the estimate of Lemma 6.3 for the Hilbert space case is better (rank $P \geq \sum_{j=1}^{\infty} \|Pe_j\|^2$) since one can use orthogonality in the proof, rather than just the triangle inequality. However, this does not affect the argument in Corollary 8.4, where we only require the rank of P to be finite.

Proof Let U be the range of P. Since P is a finite-dimensional projection, U is a finite-dimensional subspace of ℓ^p. By Exercise 7.3, U has an Auerbach basis $\{u_1, ...u_n\}$ with corresponding elements $\{f_1^*, ..., f_n^*\} \in U^*$, such that $\|u_i\|_U = 1$, $\|f_i\|_{U^*} = 1$, and $f_i^*(u_k) = \delta_{ik}$ for $1 \leq i, k \leq n$. Using the Hahn–Banach Theorem each f_i can be extended to an element $\phi_i \in \ell^q$ with $\|\phi_i\|_q = 1$ and such that $\phi_i(u_k) = \delta_{ik}$ for $1 \leq i, k \leq n$. Thus for every element $x \in \ell^p$, we can write

$$Px = \sum_{i=1}^{n} \phi_i(x) u_i.$$

It follows in particular that for each $j = 1, 2, \ldots,$

$$Pe_j = \sum_{i=1}^{n} \phi_i(e_j) u_i.$$

Now, we can expand ϕ_i using the canonical basis $\{e_j^*\}_{j=1}^{\infty}$ of ℓ^q, so that

$$\phi_i = \sum_{k=1}^{\infty} \lambda_{ik} e_k^*, \qquad \text{for every} \quad i = 1, 2, \ldots$$

with $\sum_{k=1}^{\infty} |\lambda_{ik}|^q = 1$. Thus for each $j = 1, 2, \ldots$

$$\begin{aligned}
\|Pe_j\|_{\ell^p} &= \left\| \sum_{i=1}^{n} \phi_i(e_j) u_i \right\|_{\ell^p} \\
&\leq \sum_{i=1}^{n} \|\phi_i(e_j) u_i\|_{\ell^p} = \sum_{i=1}^{n} |\phi_i(e_j)| \\
&= \sum_{i=1}^{n} \left| \sum_{k=1}^{\infty} \lambda_{ik} e_k^*(e_j) \right| = \sum_{i=1}^{n} |\lambda_{ij}| \\
&\leq n^{1/p} \left(\sum_{i=1}^{n} |\lambda_{ij}|^q \right)^{1/q}.
\end{aligned}$$

Therefore,

$$\sum_{j=1}^{\infty} \|Pe_j\|_{\ell^p}^q \leq n^{q-1} \sum_{j=1}^{\infty} \sum_{i=1}^{n} |\lambda_{ij}|^q = n^{q-1} \sum_{i=1}^{n} \left(\sum_{j=1}^{\infty} |\lambda_{ij}|^q \right) \leq n^q.$$

□

As a corollary, one can show that for the orthogonal sets introduced in Lemma 3.5, the box-counting dimension and the dual thickness coincide (for a direct argument that gives this value for the thickness in ℓ^2 and does not use

Theorem 8.1 see Lemma 7.4); from this it follows that the Hölder exponent in Theorem 8.1 is asymptotically sharp, using ℓ^2 ($p = q = 2$) for the Hilbert space case and c_0 ('$p = \infty$', $q = 1$) for the Banach space case.

Corollary 8.4 *Let $A = \{a_j e_j\}_{j=1}^{\infty} \cup \{0\}$ be an 'orthogonal' subset of ℓ^p as in Lemma 3.5. Then $\tau^*(A) = d_B(A)$ and if there exists a finite-dimensional projection P in ℓ^p and a $\theta \in (0, 1)$ such that*

$$\|\alpha\|_{\ell^p} \leq C\|P\alpha\|_{\ell^p}^{\theta}, \text{ for each } \alpha \in A, \tag{8.5}$$

then

$$\theta \leq \frac{1}{1 + (\tau^*(A)/q)}. \tag{8.6}$$

Proof Since $P(a_j e_j) = a_j P e_j$, it follows from (8.5) applied to $a_j e_j$ that

$$|a_j| \leq C|a_j|^{\theta}\|Pe_j\|_{\ell^p}^{\theta}, \qquad \text{i.e.} \quad \|Pe_j\|_{\ell^p} \geq C^{-1/\theta}|a_j|^{(1/\theta)-1}.$$

Lemma 8.3 implies that for such a P,

$$\text{rank}\, P \geq \left(\sum_{j=1}^{\infty} \|Pe_j\|_{\ell^p}^{q}\right)^{1/q} \geq C^{-\theta}\left(\sum_{j=1}^{\infty} |a_j|^{q[(1/\theta)-1]}\right)^{1/q}.$$

In particular, if P is a finite-rank projection then

$$\sum_{j=1}^{\infty} |a_j|^{q[(1/\theta)-1]} < \infty. \tag{8.7}$$

Now, using the expression for the box-counting dimension of these sequences, (3.11) from Lemma 3.5,

$$d_B(A) = \inf\{d : \sum_{n=1}^{\infty} |a_n|^d < \infty\},$$

it follows that

$$d_B(A) \leq q[(1/\theta) - 1],$$

which implies that

$$\theta \leq \frac{1}{1 + (d_B(A)/q)}. \tag{8.8}$$

We now deduce that in fact $\tau^*(A) = d_B(A)$, which will give (8.6). We know that in general $\tau^*(A) \leq d_B(A)$ (Lemma 7.8), so suppose that $\tau = \tau^*(A) < d_B(A)$. Then the result of the embedding theorem (Theorem 8.1) coupled with the Decomposition Lemma (Lemma 8.2) implies that for some k sufficiently large one can find a projection P of rank k such that (8.5) holds for some $\theta > 1/(1+(d_B(A)/q))$. But this contradicts (8.8), and so $\tau^*(A) = d_B(A)$. □

When $p = 2$, i.e. in the Hilbert space case, one can deduce that $\tau^*(A) = \tau(A) = \text{dev}(A) = d_B(A)$ for this particular class of examples.

The same argument shows that one cannot replace the upper box-counting dimension by the lower box-counting dimension and still obtain a Hölder inverse. Indeed, the sequence $\hat{H}$ defined at the end of Chapter 3 has $d_{LB}(\hat{H}) = 1/\alpha$ but $d_B(\hat{H}) = \infty$. For $p = 2$ (the Hilbert space case) the condition (8.7) becomes

$$\sum_{j=1}^{\infty} |a_j|^{2[(1/\theta)-1]} < \infty. \tag{8.9}$$

But the values of the $\{a_j\}$ used to define $\hat{H}$ are constant, of order $1/x$, for $\sim e^x$ values of j, for ever larger values of x. It follows that whatever the value of θ, the sum on the left-hand side of (8.9) diverges.

It is natural to ask how much the requirement of linearity restricts the regularity of L^{-1} that can be attained. In particular, one can ask whether it is possible to find, in general, an embedding (not necessarily linear) of X into some $\mathbb{R}^k$ that is bi-Lipschitz (i.e. L and L^{-1} are Lipschitz). In the next chapter we introduce the Assouad dimension, and show that a necessary condition for the existence of such an embedding is that the Assouad dimension of a set is finite (this condition is not sufficient, however, see Section 9.4); a simple example (Lemma 9.9) shows that there are sets with finite box-counting dimension but infinite Assouad dimension, so that a bi-Lipschitz embedding result for sets with finite box-counting dimension is not possible. A more involved example that illustrates the same thing was given by Movahedi-Lankarani (1992); again, his set is one with infinite Assouad dimension.

Exercises

8.1 Foias & Olson (1996) prove that if P_0 and P are orthogonal projections on a real Hilbert space H of equal (finite) rank and $Px = 0$ implies that $\|Px\| \le \epsilon\|x\|$ for some $\epsilon \in (0,1)$ then $\|P - P_0\| \le \epsilon$. Use this result along with Lemma 6.1 to deduce from Theorem 8.1 that if $X \subset H$, $k \in \mathbb{N}$ with $k > 2d_B(X)$, and θ satisfies (8.1) with $\alpha = 1/2$ then a dense set of rank k orthogonal projections in H are injective on X and satisfy

$$\|x - y\| < C\|P(x-y)\|^{\theta},$$

for some $C > 0$. (This is essentially the result of Foias & Olson (1996), although they do not give an explicit bound on θ.)

8.2 Suppose that $\{X_n\}_{n\in\mathbb{Z}}$ is a family of subsets of $\mathscr{B}$, such that $d_{\mathrm{B}}(X_n) \leq d$ for each $n \in \mathbb{Z}$. Show that if $k > 2d$ then there exists a linear map $L : \mathscr{B} \to \mathbb{R}^k$ and a $\theta > 0$ such that L is injective on $\bigcup_{j\in\mathbb{Z}} X_j$, and for each $n \in \mathbb{N}$ there exists a $C_n > 0$ such that

$$\|x - y\| \leq C_n |Lx - Ly|^{\theta} \qquad \text{for all} \qquad x, y \in \bigcup_{|j|\leq n} X_j.$$

[Hint: use the fact that a countable intersection of prevalent sets is prevalent.] (A version of this result is proved in Langa & Robinson (2001) for the attractors of nonautonomous systems, in Langa & Robinson (2006) for random dynamical systems, and in Robinson (2008) for general cocycle dynamical systems.)

8.3 Use Lemma 6.3 (or Lemma 8.3) to show that if $X \subset H$ contains a set of the form $\{0\} \cup \{\alpha_j\}_{j=1}^{\infty}$, where the α_j are orthogonal, then no linear map $L : H \to \mathbb{R}^k$ can be bi-Lipschitz on X. (As discussed above, in the next chapter we will see examples of sets for which no map, whether linear or not, can provide a bi-Lipschitz embedding into a Euclidean space.)

9
Assouad dimension

9.1 Homogeneous spaces and the Assouad dimension

A long-standing open problem in the theory of metric spaces is to find conditions guaranteeing that a space (X, ϱ) can be embedded in a bi-Lipschitz way into some Euclidean space (see Heinonen (2003)). The Assouad dimension was introduced in this context (Assouad, 1983; see also Bouligand (1928) for an earlier definition), and is most naturally defined as a concept auxiliary to the notion of a homogeneous set:

Definition 9.1 A subset A of a metric space (X, ϱ) is said to be (M, s)-*homogeneous* (or simply *homogeneous*) if the intersection of A with any ball of radius r can be covered by at most $M(r/\rho)^s$ balls of smaller radius ρ.

In terms of the notation used in the previous chapters, this says that

$$N(B(x, r) \cap A, \rho) \leq M(r/\rho)^s$$

for every $x \in A$ and $r > \rho$. In the light of this, it is convenient to define

$$N_A(r, \rho) = \sup_{x \in A} N(B(x, r) \cap A, \rho).$$

(Of course, if one takes $A = X$ then the '$\cap A$' is redundant.)

Lemma 9.2 *Any subset of* $\mathbb{R}^N$ *is* $(2^{N+1}, N)$*-homogeneous.*

Proof The cube $[-r, r]^N$ in $\mathbb{R}^N$ can be covered by $[(2r/\rho) + 1]^N$ cubes of side ρ. Since each cube of side $1/\rho$ lies within a ball of radius $\sqrt{N}/\rho$, the sphere of radius r (which lies within $[-r, r]^N$) can be covered by fewer than

$$[(2r/\rho) + 1]^N \leq 2^{N+1}(r/\rho)^N$$

balls of radius ρ. Clearly if X is any subset of $\mathbb{R}^N$, $X \cap B(x, r)$ can be covered by fewer than $2^{N+1}(r/\rho)^N$ balls of radius ρ. $\square$

Homogeneity is preserved under bi-Lipschitz mappings.

Lemma 9.3 *Suppose that (X, ϱ_X) is (M, s)-homogeneous and that the map $f : (X, \varrho_X) \to (Y, \varrho_Y)$ is bi-Lipschitz:*

$$L^{-1}\,\varrho_X(x_1, x_2) \le \varrho_Y(f(x_1), f(x_2)) \le L\,\varrho_X(x_1, x_2)$$

for some $L > 0$. Then $f(X)$ is an (ML^{2s}, s)-homogeneous subset of Y.

Proof Take $y \in f(X)$ and consider $B_Y(y, r) \cap f(X)$. Then $y = f(x)$ for some $x \in X$. Since f^{-1} is Lipschitz,

$$f^{-1}[B_Y(y, r) \cap f(X)] \subseteq B_X(x, Lr) \cap X.$$

Since X is (M, s)-homogeneous, $B_X(x, Lr) \cap X$ can be covered by $N \le M(Lr/(\rho/L))^s$ balls of radius ρ/L,

$$B_X(x, Lr) \cap X \subseteq \bigcup_{j=1}^{N} B_X(x_j, \rho/L).$$

Since f is Lipschitz, $f(B_X(x_j, \rho/L)) \subseteq B_Y(f(x_j), \rho)$, whence

$$B_Y(y, r) \cap f(X) \subseteq \bigcup_{j=1}^{N} f(B_X(x, Lr) \cap X) \subseteq \bigcup_{j=1}^{N} B_Y(f(x_j), \rho).$$

Hence $f(X)$ is (ML^{2s}, s)-homogeneous. □

It follows from these two elementary observations (Lemmas 9.2 and 9.3) that $A \subseteq (X, \varrho)$ must be homogeneous if it is to admit a bi-Lipschitz embedding into some $\mathbb{R}^N$. However, as we will see below, there are examples of homogeneous spaces that cannot be bi-Lipschitz embedded into any Euclidean space, so homogeneity is not sufficient for the existence of such an embedding.

A more pleasing, but equivalent, definition is that of a 'doubling' set: $A \subseteq (X, \varrho)$ is said to be *doubling* if the intersection with A of any ball of radius r can be covered by at most K balls of radius $r/2$, where K is independent of r (see Luukkainen (1998)).

Lemma 9.4 *A set $A \subseteq (X, \varrho)$ is homogeneous iff it is doubling.*

Proof That a homogeneous set is doubling is immediate. To show the converse, suppose that $N_A(r, r/2) \le K$. Given $0 < \rho < r$, choose n such that $r/2^n \le \rho < r/2^{n-1}$; then

$$N_A(r, \rho) \le N_A(r, r/2) N_A(r/2, r/4) \cdots N_A(r/2^{n-1}, r/2^n) \le K^n$$

and since $n-1 \le \log_2(r/\rho)$ it follows that

$$N_A(r,\rho) \le K(K^{n-1}) \le K(r/\rho)^{\log_2 K}. \qquad \square$$

We now define the Assouad dimension.

Definition 9.5 The *Assouad dimension* of X, $d_A(X)$, is the infimum of all s such that (X, ϱ) is (M,s)-homogeneous for some $M \ge 1$.

The following lemma gives some elementary properties of this definition. Observe that with (iv) we have now shown that for any compact set X,

$$\dim(X) \le d_H(X) \le d_{LB}(X) \le d_B(X) \le d_A(X). \tag{9.1}$$

Lemma 9.6

(i) *if $A, B \subseteq (X,\varrho)$ and $A \subseteq B$ then $d_A(A) \le d_A(B)$;*
(ii) *if $A, B \subset (X,\varrho)$ then $d_A(A \cup B) = \max(d_A(A), d_A(B))$;*
(iii) *if X is an open subset of $\mathbb{R}^N$ then $d_A(X) = N$;*
(iv) *if X is compact then $d_B(X) \le d_A(X)$; and*
(v) *d_A is invariant under bi-Lipschitz mappings.*

Proof (i) and (ii) are obvious and (v) follows from Lemma 9.3. For (iii), clearly $d_A(X) \le N$, since any subset of $\mathbb{R}^N$ is $(2^{N+1}, N)$-homogeneous (Lemma 9.2). If X is an open subset of $\mathbb{R}^N$ then it contains an open ball $B = B(x,r)$. Suppose that $d_A(X) < N$; then $d_A(B) < N$, so that B is (M,s)-homogeneous for some $s < N$. It follows that B can be covered by $M(r/\rho)^s$ balls of radius ρ, and so

$$\mu(B) \le M(r/\rho)^s \Omega_N \rho^N.$$

Since $\rho > 0$ is arbitrary and $s < N$, this implies that $\mu(B) = 0$. So $d_A(X) = N$ as claimed.

For (iv), since X is compact, $X \subset B(0,R)$ for some $R > 0$; thus for any $s > d_A(X)$,

$$N(X,\rho) = N(X \cap B(0,R), \rho) \le M(R/\rho)^s = [MR^s]\rho^{-s},$$

and hence $d_B(X) \le d_A(X)$. $\square$

In line with the characterisation of the covering dimension in terms of the Hausdorff dimension (Corollary 2.13), and the lower and upper box-counting dimensions (Exercises 3.3 and 3.4), Luukkainen (1998) proved that

$$\dim(X) = \inf\{d_A(X') : \ X' \text{ homeomorphic to } X\}.$$

Given (9.1), this shows in particular that any set X has a homeomorphic image X' such that

$$\dim(X') = d_H(X') = d_{LB}(X') = d_B(X') = d_A(X'). \tag{9.2}$$

We again repeat Luukkainen's remark that 'there is no purely topological reason for X to be fractal', which is given considerably more force by (9.2).

9.2 Assouad dimension and products

The Assouad dimension behaves like the upper box-counting dimension under the operation of taking products.

Lemma 9.7 *If (X, ϱ_X) and (Y, ϱ_Y) are metric spaces then*

$$d_A(X \times Y) \leq d_A(X) + d_A(Y),$$

where $X \times Y$ is equipped with any of the product metrics ρ_α, $1 \leq \alpha \leq \infty$, defined in (3.6) and (3.7).

Proof We use the metric

$$\varrho_\infty((x_1, y_1), (x_2, y_2)) = \max(\varrho_X(x_1, x_2), \varrho_Y(y_1, y_2))$$

on $X \times Y$ in the proof. Since this is equivalent to any of the ρ_α metrics (for $1 \leq \alpha < \infty$) and the Assouad dimension is invariant under bi-Lipschitz mappings (Lemma 9.6(v)) this will sufficient to prove the lemma.

If $d_A(X) < \alpha$ and $d_A(Y) < \beta$, there exist M_X and M_Y such that

$$N_X(r, \rho) < M_X \left(\frac{r}{\rho}\right)^\alpha \qquad \text{and} \qquad N_Y(r, \rho) < M_Y \left(\frac{r}{\rho}\right)^\beta.$$

Let B be a ball of radius r in $X \times Y$. Then $B = U \times V$, where U and V are balls of radius r in X and Y, respectively. Cover U by balls U_i of radius ρ, and the ball V by balls V_i of radius ρ. Then the products $U_i \times V_j$ form a cover of B by balls in $X \times Y$ of radius ρ, and at most

$$M_X M_Y \left(\frac{r}{\rho}\right)^{\alpha+\beta}$$

are required. It follows that $d_A(X \times Y) \leq d_A(X) + d_A(Y)$. □

The inequality here can be strict, as the following example due to Larman (1967) shows. For each $m \in \mathbb{N}$, divide the interval

$$(2^{-2^{m+1}}, 2^{-2^m})$$

into $2^{2^m} - 1$ intervals, each of length $2^{-2^{m+1}}$. Let K_m denote the collection of the $2^{2^{2m}}$ endpoints of these intervals. Set

$$A = \{0\} \cup \bigcup_{m=1}^{\infty} K_{4m} \qquad \text{and} \qquad B = \{0\} \cup \bigcup_{m=1}^{\infty} K_{4m+2}.$$

Lemma 9.8 *For the sets A and B defined above,*

$$d_A(A) = d_A(B) = d_A(A \times B) = 1.$$

Proof Clearly $d_A(A) \le 1$ and $d_A(B) \le 1$. Let $r_n = 2^{-2^{4n}}$, and consider

$$(-r_n, r_n) \cap A = \{0\} \cup \bigcup_{j=n}^{\infty} K_{4j}.$$

Then, since this contains K_{4n}, it requires at least $2^{2^{4n}} - 1$ intervals of length $\rho_n = 2^{-2^{4n+1}}$ to cover it. So

$$N_A(r_n, \rho_n) \ge 2^{2^{4n}} - 1$$

and

$$\frac{r_n}{\rho_n} = \frac{2^{-2^{4n}}}{2^{-2^{4n+1}}} = 2^{2^{4n}}.$$

So A cannot be (M, s)-homogeneous for any $s < 1$. It follows that $d_A(A) = 1$, and similarly $d_A(B) = 1$.

Now consider $A \times B$. Since $A \times B$ contains a copy of A, $d_A(A \times B) \ge 1$. Take $r = 2^{-2^{2m}}$; this is the smallest value of r such that

$$B(0, r) \cap [A \times B] = \left[\{0\} \cup \bigcup_{j=0}^{\infty} K_{2m+4j} \right] \times \left[\{0\} \cup \bigcup_{j=0}^{\infty} K_{2(m+1)+4j} \right].$$

The number of balls required in a cover by ρ-balls is essentially determined by the value of t for which $B(0, \rho) \supset K_t$. So choose $\rho = 2^{-2^{2n+1}}$; the largest value of ρ such that K_t is in $B(0, \rho)$ for every $t > 2n$ but not for $t = 2n$.

Then the number of balls required to cover $B(0, r) \cap [A \times B]$ is bounded by

$$N \le (n - m + 1)^2 \times 2^{2^{2n}} \times 2^{2^{2n-2}},$$

while

$$\frac{r}{\rho} = 2^{2^{2n+1} - 2^{2m}}.$$

It follows that $N \le (r/\rho)^s$, where

$$s \le \frac{2 \log_2(n - m + 1) + 2^{2n} + 2^{2n-2}}{2^{2n+1} - 2^{2m}} \le 1. \qquad \square$$

9.3 Orthogonal sequences

We now investigate further some surprising properties of the Assouad dimension, following Olson (2002). We have used the example of orthogonal sequences in a Hilbert space in previous chapters, and again this simple class of examples is illuminating.

Lemma 9.9 *Let $X = \{n^{-\alpha} e_n\} \cup \{0\}$, where $\{e_n\}_{n=1}^{\infty}$ is an orthonormal subset of a Hilbert space H. Then $d_A(X) = \infty$.*

(Recall that $d_H(X) = 0$ and that $d_B(X) = 1/\alpha$, see (3.13).)

Proof Let $r_m = m^{-\alpha}$, and consider

$$B(0, r_m) \cap X = \{n^{-\alpha} e_n \, : \, n \geq m\} \cup \{0\}.$$

Now cover $B(0, r_m) \cap X$ by balls of radius $r_m/2$; each point in this set of norm more than $r_m/2$ will require its own ball, and since

$$n^{-\alpha} > r_m/2 \qquad \Rightarrow \qquad n < (2/r_m)^{1/\alpha} = m2^{1/\alpha}$$

it follows that $N(B(0, r_m) \cap X, r_m/2) \geq m2^{1/\alpha} - m - 1$. Since the right-hand side tends to infinity as $m \to \infty$, X cannot be doubling. □

As remarked at the end of the previous chapter, this gives a simple example of a compact set with finite box-counting dimension that cannot be bi-Lipschitz embedded into any $\mathbb{R}^k$.

However, a geometric sequence has zero Assouad dimension, as the next (more general) result shows.

Lemma 9.10 *Let $\{e_n\}_{n=1}^{\infty}$ be the canonical basis of ℓ^p, $1 \leq p < \infty$, or of c_0, and consider the set $X = \{a_n e_n\}_{n=1}^{\infty} \cup \{0\}$. Suppose that there exist K and α with $K > 0$ and $0 < \alpha < 1$ such that*

$$K^{-1}\alpha^n \leq a_n \leq K\alpha^n.$$

Then $d_A(X) = 0$, where the dimension is taken in ℓ^p $(1 \leq p < \infty)$ or c_0.

Proof Take $0 < \rho < r$. Consider a ball of radius r centred at the origin; then

$$\begin{aligned} B(0, r) \cap X &= \{a_n e_n \, : \, a_n < r\} \cup \{0\} \\ &\subseteq \{a_n e_n \, : \, K\alpha^n < r\} \cup \{0\} \\ &= \{a_n e_n \, : \, n > (\log r - \log K)/\log \alpha\} \cup \{0\} \\ &\subseteq \{a_n e_n \, : \, n \geq [(\log r - \log K)/\log \alpha] - 1\} \cup \{0\}. \end{aligned}$$

A cover of $B(0, r) \cap X$ by balls of radius ρ will require a separate ball for each point of norm greater than ρ; since

$$K^{-1}\alpha^n > \rho \qquad \Rightarrow \qquad a_n > \rho,$$

it follows that $a_n > \rho$ for $n < (\log \rho + \log K)/\log \alpha$, so certainly the same is true for $n \leq [(\log \rho + \log K)/\log \alpha] + 1$. Thus

$$\begin{aligned} N(B(0, r) \cap X, \rho) &\leq \frac{\log \rho + \log K}{\log \alpha} - \frac{\log r - \log K}{\log \alpha} + 2 \\ &= \frac{1}{-\log \alpha} \log\left(\frac{r}{\rho}\right) + \frac{2 \log K}{\log \alpha} + 2. \end{aligned}$$

So X is (M, s)-homogeneous for any $s > 0$, i.e. $d_{\rm A}(X) = 0$. □

The lower bound in this result is necessary:

Lemma 9.11 *There are sequences a_n converging arbitrarily fast to zero for which*

$$X = \{a_n e_n\}_{n=1}^{\infty} \cup \{0\}$$

has $d_{\rm A}(X) = \infty$.

Proof Let b_j be a sequence that converges to zero. Let $a_n = b_j$ for $2^{j-1} \leq n \leq 2^j - 1$. Now let $r = b_j + \epsilon$ and consider

$$B(0, r) \cap X = \{a_n e_n : n \geq 2^{j-1}\}.$$

The number of balls of radius $r/2$ required to cover $B(0, r) \cap X$ is larger than $2^j - 2^{j-1}$, which is unbounded as $j \to 0$. So X is not doubling. Since b_j can converge arbitrarily fast to zero, so can a_n. □

Assumptions on the set of differences $X - X$ are the key to proving embedding results that use linear maps. We have seen that while for the box-counting dimension $d_{\rm B}(X - X) \leq 2d_{\rm B}(X)$, one can have sets with zero Hausdorff dimension for which $d_{\rm H}(X - X) = \infty$. Unfortunately the same is true of the Assouad dimension.

Lemma 9.12 *There exists a set X with $d_{\rm A}(X) = 0$ and $d_{\rm A}(X - X) = \infty$.*

Proof Let $\{x_j\}$ be an orthogonal sequence of the type constructed in the previous lemma, with $\|x_j\| \leq 4^{-j}$. Suppose that the complement of the linear span of the $\{x_j\}$ is infinite-dimensional, and choose a second orthogonal sequence $\{y_j\}$ in this complement with $\|y_j\| = 4^{-j}$.

Let X be the closure of the set $\{a_j\}$, where

$$a_{2j} = y_j \qquad \text{and} \qquad a_{2j+1} = x_j + y_j.$$

Clearly $X - X$ contains $\{x_j\}$, and so $d_A(X - X) = \infty$. However, Lemma 9.10 implies that $d_A(X) = 0$: for $k = 2j$,

$$\|a_k\| = \|y_j\| = 4^{-j} = 2^{-k},$$

while for $k = 2j + 1$,

$$\|a_k\| \leq \|x_j\| + \|y_j\| \leq 4^{-j} + 4^{-j} = 2(4^{-(k-1)/2}) = 4 \times 2^{-k}$$

and

$$\|a_k\| \geq \|y_j\| = 2 \times 2^{-k}. \qquad \square$$

Note that in a very roundabout way we have shown that the Assouad dimension can increase under Lipschitz continuous transformations, since $d_A(X \times X) \leq 2d_A(X)$, and $X - X$ is the image of $X \times X$ under the Lipschitz mapping $(x, y) \mapsto x - y$.

The following result, again due to Olson (see Olson & Robinson (2010)) is more positive, and will be useful below.

Lemma 9.13 *Let $X = \{x_j\}_{j=1}^{\infty}$ be an orthogonal sequence in H. If $d_A(X) < \infty$ then $d_A(X - X) \leq 2d_A(X)$.*

Proof Suppose that X is (M, s)-homogeneous. Set $B_X(r, x) = X \cap B(r, x)$, and consider a ball $B = B_{X-X}(r, x - y) \subseteq X - X$ of radius r that is centred at $x - y \in X - X$. Since $B \subseteq B_{X-X}(\rho, 0) \cup \big(B \setminus \{0\}\big)$, we need only cover $B \setminus \{0\}$.

Suppose that $x = y$, so that $B = B_{X-X}(r, 0)$. Let $a - b \in B \setminus \{0\}$. Then $a \neq b$ and therefore a is orthogonal to b. It follows that

$$\big\|(a - b) - (x - y)\big\|^2 = \|a\|^2 + \|b\|^2 < r^2.$$

Hence $a, b \in B_X(r, 0)$, and consequently

$$B \setminus \{0\} \subseteq B_X(r, 0) - B_X(r, 0).$$

Cover $B_X(r, 0)$ with $M(2r/\rho)^s$ balls $B_X(\rho/2, a_i)$ of radius $\rho/2$ centred at $a_i \in X$. Then

$$\begin{aligned}\bigcup_{i,j} B_{X-X}(\rho, a_i - a_j) &\supseteq \bigcup_i B_X(\rho/2, a_i) - \bigcup_j B_X(\rho/2, a_j) \\ &\supseteq B_X(r, 0) - B_X(r, 0) \supseteq B_{X-X}(r, 0) \setminus \{0\}.\end{aligned}$$

It follows that B is covered by $1 + M^2(2r/\rho)^{2s}$ balls of radius ρ.

Now suppose that $x \neq y$. Let $a - b \in B\backslash\{0\}$. Again $a \neq b$ and therefore a is orthogonal to b. Therefore

$$\|(a-b)-(x-y)\|^2 = \begin{cases} \|a-x\|^2 + \|b-y\|^2 & \\ \|a+y\|^2 + \|2x\|^2 & \text{if} \\ \|2y\|^2 + \|b+x\|^2 & \end{cases} \begin{matrix} a \neq y, \ b \neq x \\ a \neq y, \ b = x \\ a = y, \ b \neq x \end{matrix},$$

and so

$$\left.\begin{matrix} a \in B_X(r,x) & b \in B_X(r,y) \\ a \in B_X(r,-y) & b \in B_X(r,x) \\ a \in B_X(r,y) & b \in B_X(r,-x) \\ a \in B_X(r,y) & b \in B_X(r,x) \end{matrix}\right\} \text{ if } \begin{cases} a \neq y, \ b \neq x \\ a \neq y, \ b = x \\ a = y, \ b \neq x \\ a = y, \ b = x \end{cases}.$$

Therefore

$$\begin{aligned} B \setminus \{0\} \subseteq \big(B_X(r,x) - B_X(r,y)\big) &\cup \big(B_X(r,-y) - B_X(r,x)\big) \\ \cup\big(B_X(r,y) - B_X(r,-x)\big) &\cup \big(B_X(r,y) - B_X(r,x)\big). \end{aligned}$$

Cover each of $B_X(r,x)$, $B_X(r,-x)$, $B_X(r,y)$, and $B_X(r,-y)$ by $M(2r/\rho)^s$ balls of radius $\rho/2$. An argument similar to that used before yields a cover of B by $1 + 4M^2(2r/\rho)^{2s}$ balls of radius $r/2$.

Since $N_{X-X}(r,\rho) \leq 1 + 4M^2(2r/\rho)^{2s}$ it follows that $d_A(X - X) \leq 2s$. □

9.4 Homogeneity is not sufficient for a bi-Lipschitz embedding

We have already remarked that any set that can be bi-Lipschitz embedded into $\mathbb{R}^k$ must be homogeneous, but the following example, due to Lang & Plaut (2001; after Laakso (2002)) shows that this is not sufficient.

The construction yields a metric space that is doubling but cannot be bi-Lipschitz embedded into any Hilbert space (finite- or infinite-dimensional). This example is somewhat simpler than the 'classical' example of the Heisenberg group equipped with the Carnot–Carathéodory metric (see Semmes (1996), for example).

Let X_0 be $[0, 1]$ with the standard Euclidean distance. To construct X_{i+1} from X_i, take six copies of X_i and rescale by a factor of $\frac{1}{4}$. Arrange four in a 'square' by identifying pairs of endpoints, and then attach the remaining two copies at 'opposite' points, see Figure 9.1.

At every step X_i has diameter 1, has two endpoints, and consists of 6^i edges each of which has length 4^{-i}. The metric $\varrho_i(x, y)$ on X_i is the geodesic

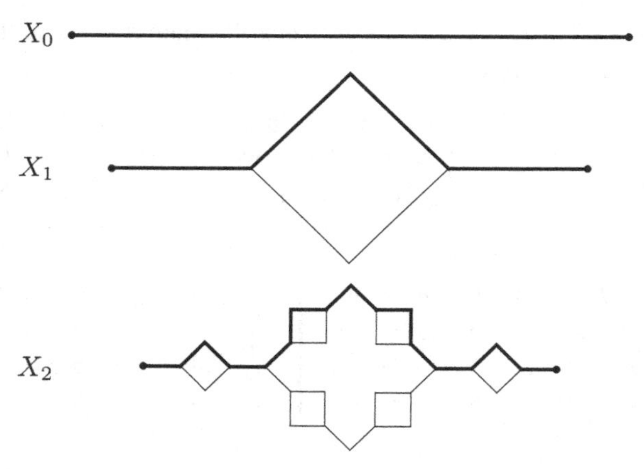

Figure 9.1 The first steps of the construction of the geodesic metric space (X, ϱ). At each stage the bold subset is isometric to X_0.

distance: the shortest distance that one needs to travel on the graph from x to y. For every $j > i$, (X_j, ϱ_j) contains an isometric copy of (X_i, ϱ_i), and $\mathrm{dist}(X_j, X_i) < (1/4)^{i+1}$ (see Figure 9.1) and so $\{(X_i, \varrho_i)\}_{i=1}^{\infty}$ forms a Cauchy sequence in the Gromov–Hausdorff metric[1] (see Chapter 3 of Gromov (1999), or Heinonen (2003)). It follows that this sequence converges to some limiting compact metric space (X, ϱ); there remain isometric copies of (X_i, ϱ_i) in (X, ϱ). The exact details of this limiting argument are not necessary here, the key point is that this process leads to such a limit set containing isometric copies of every (X_i, ϱ_i).

Lemma 9.14 *The space (X, ϱ) is doubling with doubling constant* 6, *and if H is a Hilbert space and $f : X_i \to H$ satisfies $\|f(x) - f(y)\| \geq \varrho(x, y)$ then the Lipschitz constant of f is bounded below by $(1 + (i/4))^{1/2}$. In particular, there is no bi-Lipschitz embedding of X into a Hilbert space.*

Proof Take $x \in X$ and r with $0 < r \leq \frac{1}{2}$. Choose i with

$$\frac{r}{2} \leq \left(\frac{1}{4}\right)^i < 2r,$$

[1] First, given two subsets A, B of a metric space (X, ϱ), define the symmetric Hausdorff distance $\mathrm{dist_H}(A, B) = \max(\mathrm{dist}(A, B), \mathrm{dist}(B, A))$. The Gromov–Hausdorff distance between two metric spaces A and B is defined as

$$d_{\mathrm{GH}}(A, B) = \inf_{A', B' \in M} \mathrm{dist_H}(A', B'),$$

where the infimum is taken over all isometric images A' and B' of A and B as subsets of ℓ^∞ (at least one such isometry always exists, see Exercise 9.2).

and let $h : X_i \to X$ be an isometric embedding of X_i into X with $x = h(x')$ for some $x' \in X_i$. Let

$$Z = \partial B(x', r) \cup (B(x', r) \cap \{p, q\}),$$

where p and q are the endpoints of X_i. Then since $r \leq 2(4^{-i})$, Z certainly contains no more than six points, and the closed balls in X_i of radius r centred at the points of Z cover $B(x', 2r)$. Then the closed balls of radius r centred at the points of $h(Z)$ also cover $B(x, 2r)$, because the edge cycles generated after the ith step have length no larger than $4(4^{-(i+1)}) < 2r$: if z lies on such a cycle, $B(z, r)$ covers the whole cycle. It follows that X is doubling with constant 6.

We show by induction that for any $f : X_i \to H$ as in the statement of the lemma, there exist two consecutive vertices x and x' such that

$$\|f(x) - f(x')\|^2 \geq \left(1 + \frac{i}{4}\right) \varrho\left(x, x'\right)^2.$$

This is certainly true for $i = 0$. Take $i = k \geq 1$ and assume that the result is true for $i = k - 1$. Since X_k contains an isometric copy of X_{k-1}, there exist points $x_0, x_2 \in X_k$ corresponding to two adjacent vertices of X_{k-1} such that

$$\|f(x_0) - f(x_2)\|^2 \geq \left(1 + \frac{k-1}{4}\right) \varrho(x_0, x_2)^2. \tag{9.3}$$

Let x_1 and x_3 be the two midpoints between x_0 and x_2; then $\varrho(x_1, x_3) = \frac{1}{2}\varrho(x_0, x_2)$. Setting $x_4 = x_0$ we have

$$\|f(x_0) - f(x_2)\|^2 + \|f(x_1) - f(x_3)\|^2 \leq \sum_{j=0}^{3} \|f(x_j) - f(x_{j+1})\|^2,$$

this 'quadrilateral inequality' holding in any inner product space, see Exercise 9.1. Since

$$\begin{aligned}\|f(x_0) - f(x_2)\|^2 + \|f(x_1) - f(x_3)\|^2 &\geq \left(1 + \frac{k-1}{4}\right) \varrho(x_0, x_2)^2 + \varrho(x_1, x_3)^2 \\ &= \left(1 + \frac{k}{4}\right) \varrho(x_0, x_2)^2,\end{aligned}$$

it follows that for some $j \in \{0, \ldots, 3\}$

$$\|f(x_j) - f(x_{j+1})\|^2 \geq \frac{1}{4}\left(1 + \frac{i}{4}\right) \varrho(x_0, x_2)^2 = \left(1 + \frac{k}{4}\right) \varrho(x_j, x_{j+1})^2.$$

Now take $x = x_j$ and x' to be one of the midpoints between x_j and x_{j+1} to obtain (9.3) for $i = k$.

Now suppose that $f : X \to H$ is an embedding of X into H with

$$L^{-1}\varrho(x, y) \le \|f(x) - f(y)\| \le L\varrho(x, y).$$

Setting $g(x) = Lf(x)$ gives a $g : (X, \varrho) \to H$ such that

$$\varrho(x, y) \le \|g(x) - g(y)\| \le L^2\varrho(x, y).$$

But X contains an isometric copy of every X_i, which implies that $\mathrm{Lip}(g) \ge 1 + (i/4)$ for all i, a contradiction. □

The strongest result for sets with finite Assouad dimension is due to Assouad (1983), who showed that any metric space with $d_A(X) < \infty$ can be mapped via $\phi : X \to \mathbb{R}^k$ into some finite-dimensional Euclidean space in a bi-Hölder way,

$$\frac{1}{c}\,\varrho(s, t)^\alpha \le |\phi(s) - \phi(t)| \le c\,\varrho(s, t)^\alpha$$

for all $0 < \alpha < 1$; and that this characterises sets with finite Assouad dimension. Olson & Robinson (2010) showed that such sets can be mapped in an almost bi-Lipschitz way into an infinite-dimensional Hilbert space: for every $\gamma > \frac{1}{2}$ there exists a map $f : (X, \varrho) \to H$ such that for some $L > 0$

$$\frac{1}{L}\frac{\varrho(x, y)}{(\operatorname{slog}\varrho(x, y))^\gamma} \le \|f(x) - f(y)\| \le L\,\varrho(x, y),$$

where $\operatorname{slog}(x) = \log(x + x^{-1})$ (cf. Proposition 7.18 in Benyamini & Lindenstrauss (2000)).

We will show in the next section that if $X - X$ is a homogeneous compact subset of a Banach space, then one can find an almost bi-Lipschitz embedding into some Euclidean space, i.e. an embedding that is bi-Lipschitz to within logarithmic corrections.

9.5 Almost bi-Lipschitz embeddings

As with the embedding theorems for the Hausdorff and box-counting dimension, we construct a probe space that is tailored to the particular set (and notion of dimension) that we are considering. The following simple results (Lemmas 9.15 and 9.16) are the key to the argument used to prove Theorem 9.18.

Lemma 9.15 *Suppose that Z is a compact homogeneous subset of a Banach space $\mathscr{B}$. Then there exists an $M' > 0$ and a sequence of linear subspaces $\{V_j\}_{j=0}^\infty$ of $\mathscr{B}^*$ with $\dim V_j \le M'$ for every j, such that for any $z \in Z$ with $2^{-(n+1)} \le \|z\| \le 2^{-n}$, there exists an element $\psi \in V_n$ such that*

$$\|\psi\|_* = 1 \qquad \text{and} \qquad |\psi(z)| \ge 2^{-(n+3)}. \tag{9.4}$$

Proof Write

$$\Delta_j = \{z \in Z : 2^{-(j+1)} \leq \|z\| \leq 2^{-j}\}.$$

Since $\Delta_j \subset B(0, 2^{-j})$ it can be covered by M_j balls of radius $2^{-(j+3)}$, with the centres $\{u_i^{(j)}\}_{i=1}^{M_j}$ of these balls satisfying $\|u_i^{(j)}\| \geq 2^{-(j+2)}$, where

$$M_j = N_X(2^{-j}, 2^{-(j+3)}) \leq 8^s M = M'.$$

For each of the points $u_i^{(j)}$, use the Hahn–Banach Theorem to find a norm 1 element $\phi_i^{(j)}$ of $\mathscr{B}^*$ such that

$$\phi_i^{(j)}(u_i^{(j)}) = \|u_i^{(j)}\|.$$

For each $n \geq 0$ let V_j be the subspace of $\mathscr{B}^*$ spanned by $\{\phi_1^{(j)}, \phi_2^{(j)}, \ldots, \phi_{M_j}^{(j)}\}$. By the above, $\dim V_j \leq M'$ for all $j \geq 0$.

For any $z \in \Delta_n$ there exists a $u = u_i^{(n)}$ such that

$$\|z - u\| < 2^{-(n+3)}.$$

Writing ϕ for $\phi_i^{(n)} \in V_n$,

$$\begin{aligned}|\phi(x)| = |\phi(u) - \phi(u - z)| &\geq \|u\| - \|u - z\| \\ &\geq 2^{-(n+2)} - 2^{-(n+3)} = 2^{-(n+3)},\end{aligned}$$

and the lemma follows. □

In a Hilbert space it is more helpful to use the following result; note that the spaces V_j are now mutually orthogonal, but that the space V_n alone is not sufficiently 'rich' to obtain (9.5) (cf. (9.4), where $\psi \in V_n$ is enough).

Lemma 9.16 *Suppose that Z is a compact homogeneous subset of a Hilbert space H. Then there exists an $M' > 0$ and a sequence $\{V_j\}_{j=0}^{\infty}$ of mutually orthogonal linear subspaces of H, with $\dim V_j \leq M'$ for every j, such that for any $z \in Z$ with $2^{-(j+1)} \leq \|z\| \leq 2^{-j}$,*

$$\|\Pi_j z\| \geq 2^{-(j+2)}, \tag{9.5}$$

where Π_j is the orthogonal projection onto $\oplus_{i=1}^{j} V_i$.

Proof Write

$$Z_j = \{z \in Z : 2^{-(j+1)} \leq \|z\| \leq 2^{-j}\}.$$

Since $Z_j \subset B(0, 2^{-j})$ it can be covered by M_j balls of radius $2^{-(j+2)}$, with centres $\{u_i^{(j)}\}_{i=1}^{M_j}$, where

$$M_j = N(2^{-j}, 2^{-(j+2)}) \leq 4^s M = M'.$$

Let U_j be the space spanned by $\{u_i^{(j)}\}_{i=1}^{M_j}$; clearly $\dim(U_j) \leq M'$, and if P_j denotes the projection onto U_j,

$$\|P_j z\| \geq \|z\| - \|z - P_j z\| \geq 2^{-(j+1)} - 2^{-(j+2)} = 2^{-(j+2)}.$$

Finally, define mutually orthogonal subspaces V_j such that

$$\bigoplus_{j=1}^{n} V_j = \bigoplus_{j=1}^{n} U_j$$

and the result follows since $\|\Pi_n z\| \geq \|P_n z\|$. □

The spaces whose existence is guaranteed by these two Lemmas with $Z = X - X$ form the basis of the construction of the 'probe space' with respect to which it will be shown that linear embeddings with log-Lipschitz inverses are prevalent.

We now construct the probe space E and the associated measure μ: in the Banach space case we take $\gamma > 1$ and follow the standard construction of Section 5.2.2, but in the Hilbert space case we follow the alternative construction outlined at the end of Section 5.2.1, capitalising on the fact that the spaces $\{V_j\}$ are mutually orthogonal so that we can take $\gamma > 1/2$, and that $\dim V_j \leq M'$ for all j so that we can build our probe space from products of cubes rather than products of spheres.

The following bound, a consequence of the estimates in Section 5.2, is central to the proof. (Note that in the Banach space case we only require Lemma 5.10, but in the Hilbert space case we need the somewhat more subtle result of Lemma 5.8.)

Lemma 9.17 *If* $z \in Z$ *with* $2^{-(j+1)} \leq \|z\| \leq 2^{-j}$ *then for any* $f \in \mathscr{L}(\mathscr{B}, \mathbb{R}^N)$,

$$\mu\{L \in E : |(f + L)z| < \epsilon 2^{-j}\} \leq C\epsilon^N j^{sN}, \tag{9.6}$$

where $C = C(N)$.

Proof In the Banach space case, Lemma 5.10 guarantees that for any $\psi \in S_j$,

$$\mu\{L \in E : |(f + L)z| < \epsilon\} \leq \left(j^s d_j \epsilon |\psi(z)|^{-1}\right)^N.$$

In the case considered here, $d_j = \dim(V_j) \leq M'$, and using Lemma 9.15 there exists a $\psi \in S_j$ with $\|\psi\|_* = 1$ such that $\psi(z) \geq 2^{-(j+3)}$, from which (9.6) follows immediately. In the Hilbert space case, Lemma 5.8 ensures that

$$\mu\{L \in E : |(L + f)(x)| < \epsilon\} \leq c \left(j^s (M')^{1/2} \frac{\epsilon}{\|\Pi_j x\|}\right)^k,$$

where Π_j is the orthogonal projection onto $V_1 \oplus V_2 \oplus \cdots \oplus V_j$, and (9.6) follows using Lemma 9.16. □

Olson (2002) proved a version of the following theorem for subsets of $\mathbb{R}^N$, and Olson & Robinson (2010) gave a proof for subsets of a Hilbert space that yields $\gamma > 3/2$ as $N \to \infty$. Robinson (2009) used the probe space construction of Section 5.2.2 to prove the result for subsets of a Banach space with $\gamma > 2$. The reduction to the optimal exponents here, whose possibility was strongly suggested by the analysis in Pinto de Moura & Robinson (2010a) – see Section 9.6 – is due to Robinson (2010).

Theorem 9.18 *Let X be a compact subset of a real Banach space $\mathscr{B}$ such that $d_A(X - X) < s < N$, where $N \in \mathbb{N}$. If*

$$\gamma > \frac{\alpha N + 1}{N - s}, \tag{9.7}$$

where $\alpha = 1/2$ if $\mathscr{B}$ is a Hilbert space and $\alpha = 1$ if $\mathscr{B}$ is a general Banach space, then a prevalent set of linear maps $L : \mathscr{B} \to \mathbb{R}^N$ are injective on X and, in particular, γ-almost bi-Lipschitz: for some constant $c_L > 0$ and $\rho_L > 0$

$$\frac{1}{c_L} \frac{\|x - y\|}{|\log \|x - y\||^\gamma} \leq |Lx - Ly| \leq c_L \|x - y\| \tag{9.8}$$

for all $x, y \in X$ with $\|x - y\| < \rho_L$.

Proof Choose $\zeta > \alpha$ small enough to ensure that

$$\gamma > \frac{\zeta N + 1}{N - s}. \tag{9.9}$$

Let $\mathscr{V} = \{V_n\}_{n=1}^{\infty}$, where V_n are the spaces whose existence is guaranteed by Lemma 9.15, and set $E = E_\zeta(\mathscr{V})$ and $\mu = \mu_\zeta(\mathscr{V})$ following the construction of Section 5.2.

Denote by S_1 the set of all those $L \in \mathscr{L}(\mathscr{B}, \mathbb{R}^N)$ for which (9.8) holds for all $x, y \in X$ with $\|x - y\| < \rho_L$ for some $\rho_L > 0$. First we show that S_1 is prevalent, and then combine this with the result of Theorem 8.1 to deduce the result as stated.

Given $f \in \mathscr{L}(\mathscr{B}, \mathbb{R}^N)$, let K be a Lipschitz constant valid for all $f + L$ with $L \in E$. Define a sequence of layers of $X - X$,

$$Z_j = \{z \in X - X : 2^{-(j+1)} \leq \|z\| \leq 2^{-j}\}$$

and the corresponding set of maps that fail to satisfy the almost bi-Lipschitz property for some $z \in Z_j$,

$$Q_j = \{L \in Q : |(f + L)(z)| \leq j^{-\gamma} 2^{-j} \text{ for some } z \in Z_j\}.$$

By assumption $d_A(X - X) < s$, and so $Z_j \subset B(0, 2^{-j})$ can be covered by $N_j \le Mj^{\gamma s}$ balls of radius $j^{-\gamma}2^{-j}$. Let the centres of these balls be $z_i^{(j)} \in Z_j$ where $i = 1, \ldots, N_j$. Given any $z \in Z_j$ there is $z_i^{(j)}$ such that

$$\|z - z_i^{(j)}\| \le j^{-\gamma}2^{-j}.$$

Thus

$$\begin{aligned} |(f + L)(z)| &\ge |(f + L)(z_i^{(j)})| - |(f + L)(z - z_i^{(j)})| \\ &\ge |(f + L)(z_i^{(j)})| - Kj^{-\gamma}2^{-j}, \end{aligned}$$

which implies, using Lemma 9.17, that

$$\begin{aligned} \mu(Q_j) &\le \sum_{i=1}^{N_j} \mu\{L \in Q : |(f + L)(z_i^{(j)})| \le (1 + K)j^{-\gamma}2^{-j}\} \\ &\le N_j\big(d_j^\alpha j^\zeta (1 + K)j^{-\gamma}2^{-j}|\psi(z_i^{(j)})|^{-1}\big)^N \end{aligned}$$

for any $\psi \in V_j$.

Lemma 9.15 implies that there exits a $\psi \in V_j$ such that $|\psi(z_i^{(j)})| \ge 2^{-(j+3)}$, and since $N_j \le Mj^{\gamma s}$ and $d_j \le M'$,

$$\mu(Q_j) \le c\, j^{\gamma s}\big((M')^\alpha\, j^\zeta\, j^{-\gamma}\big)^N = c\, j^{\gamma s + N(\zeta - \gamma)}.$$

Since (9.9) implies that $\gamma s + N(\zeta - \gamma) < -1$, it follows that

$$\sum_{j=1}^{\infty} \mu(Q_j) < \infty.$$

Using the Borel–Cantelli Lemma (Lemma 4.2), μ-almost every L is contained in only a finite number of the Q_j: thus for μ-almost every L there exists a j_L such that for all $j \ge j_L$,

$$2^{-(j+1)} \le \|z\| \le 2^{-j} \quad \Rightarrow \quad |(f + L)(z)| \ge j^{-\gamma}2^{-j},$$

so for $\|z\| \le 2^{-j_L}$,

$$|(f + L)(z)| \ge 2^{-(1+\gamma)} \frac{\|z\|}{|\log \|z\||^\gamma}. \tag{9.10}$$

So the set S_1 is prevalent as claimed.

Now, since $\tau^*(X) \le d_B(X) \le d_B(X - X) \le d_A(X - X)$ we can apply Theorem 8.1 to obtain a prevalent set S_2 of linear functions $f : \mathscr{B} \to \mathbb{R}^N$ that are injective on X. Since the intersection of prevalent sets is prevalent (Corollary 5.4), there exists a prevalent set of linear maps that are injective on X and satisfy (9.10). □

9.6 Sharpness of the logarithmic exponent

We can once more use an appropriate choice of orthogonal sequence to show, following Pinto de Moura & Robinson (2010a), that the bound on the logarithmic exponent in Theorem 9.18 is asymptotically sharp: as $N \to \infty$, we obtain from (9.7) that it is possible to embed into some $\mathbb{R}^N$ with any logarithmic exponent $\gamma > 1$ in the Banach space case and $\gamma > 1/2$ in the Hilbert space case.

We consider the 'orthogonal set' $A = \{\mathrm{e}^{-n} e_n\}_{n=1}^{\infty} \cup \{0\}$, where $\{e_n\}$ is the canonical basis of ℓ^p $(1 \le p < \infty)$ or c_0. Lemma 9.10 guarantees that as a subset of ℓ^p $(1 \le p < \infty)$ and c_0 this set has zero Assouad dimension. Lemma 9.13, which extends to subsets of ℓ^p of this form, guarantees that $d_{\mathrm{A}}(A - A) = 0$.

Suppose that there exists a linear map $L : \ell^p \to \mathbb{R}^N$ and a $\rho_L > 0$ such that for every $x, y \in A$ with $\|x - y\| < \rho_L$,

$$\frac{1}{c_L} \frac{\|x - y\|}{|\log \|x - y\||^{\gamma}} \le |Lx - Ly| \le c_L \|x - y\| \tag{9.11}$$

for some $c_L > 0$. Then the Decomposition Lemma (Lemma 8.2) ensures that there exists a rank N projection P that satisfies a lower bound of the same form,

$$\|Px - Py\| \ge c \frac{\|x - y\|}{|\log \|x - y\||^{\gamma}}$$

for all $x, y \in A$ with $\|x - y\| < \rho_L$.

In particular, since $0 \in A$, one must have

$$\|Px\| \ge c \frac{\|x\|}{|\log \|x\||^{\gamma}}$$

for every $x \in A$ with $\|x\| < \rho_L$: for $x = \mathrm{e}^{-j} e_j$ this gives

$$\mathrm{e}^{-j} \|Pe_j\| \ge c \frac{\mathrm{e}^{-j} \|e_j\|}{j^{\gamma}} \qquad \Rightarrow \qquad \|Pe_j\| \ge c j^{-\gamma}.$$

Using Lemma 8.3 it follows that

$$N \ge \left(\sum_{j=1}^{\infty} \|Pe_j\|^q \right)^{1/q} \ge c \left(\sum_{j=1}^{\infty} j^{-q\gamma} \right)^{1/q}.$$

Thus to allow the existence of a finite-rank linear map satisfying (9.11) we need the sum on the right-hand side to converge, i.e. we must have $\gamma > 1/q$. In the Hilbert space case ($p = 2$) this shows that we cannot improve on $\gamma > 1/2$, while in the Banach space case (taking '$p = \infty$', i.e. c_0) we cannot improve on $\gamma > 1$.

9.7 Consequences for embedding compact metric spaces

A result valid for Banach spaces can be converted into a result for metric spaces via the Kuratowski isometric embedding of (X, ϱ) into $L^\infty(X)$:

Lemma 9.19 *Let (X, ϱ) be a compact metric space. Then the mapping $\mathscr{F} : (X, \varrho) \to L^\infty(X)$ given by $x \mapsto \varrho(x, \cdot)$ is an isometry.*

If X is not compact one can obtain the same result by choosing any $a \in X$ and considering $x \mapsto \varrho(x, \cdot) - \varrho(a, \cdot)$.

Proof Since (X, ϱ) is compact it is bounded, so $|\varrho(x, y)| \leq \mathrm{diam}(X)$ for every $x, y \in X$, i.e. $\|\mathscr{F}x\|_\infty \leq \mathrm{diam}(X)$, so $\mathscr{F}x \in L^\infty(X)$. To show that $\mathscr{F}$ is an isometry, note that by the triangle inequality

$$|(\mathscr{F}x_1)(y) - (\mathscr{F}x_2)(y)| = |\varrho(x_1, y) - \varrho(x_2, y)| \leq \varrho(x_1, x_2)$$

and

$$|(\mathscr{F}x_1)(x_1) - (\mathscr{F}x_2)(x_1)| = \varrho(x_1, x_2),$$

and so

$$\|\mathscr{F}x_1 - \mathscr{F}x_2\|_\infty = \varrho(x_1, x_2). \qquad \square$$

In this way one can interpret '$X - X$' for an arbitrary metric space (X, ϱ), i.e.

$$X - X = \{f \in L^\infty(X) : \ f = \varrho(x, \cdot) - \varrho(y, \cdot), \ x, y \in X\}. \tag{9.12}$$

The following result is then an immediate corollary of Theorem 9.18.

Theorem 9.20 *Let (X, ϱ) be a compact metric space such that $X - X$ as defined in (9.12) is a homogeneous subset of $L^\infty(X)$. Then for some $N \in \mathbb{N}$ there exists an injective almost bi-Lipschitz map $f : (X, \varrho) \to \mathbb{R}^N$.*

Exercises

9.1 Show that if x, y, z are elements of any inner product space then

$$\left\| x - \frac{y + z}{2} \right\|^2 \leq \frac{1}{2}\|x - y\|^2 + \frac{1}{2}\|x - z\|^2 - \frac{1}{4}\|y - z\|^2, \tag{9.13}$$

and deduce that

$$\|x_0 - x_2\|^2 + \|x_1 - x_3\|^2 \leq \sum_{j=0}^{3} \|x_j - x_{j+1}\|^2 \tag{9.14}$$

where $x_4 = x_0$. [Hint: consider the triples $\{x_0, x_1, x_3\}$ and $\{x_2, x_1, x_3\}$.]

9.2 Let (X, ϱ) be a separable metric space. If $\{x_j\}_{j=0}^{\infty}$ is a countable dense subset of X, show that the map $x \mapsto s(x)$ with

$$s_j(x) = \varrho(x, x_j) - \varrho(x_j, x_0) \qquad j = 1, 2, \ldots$$

provides an isometric embedding of (X, ϱ) into ℓ^∞. (Unlike the result of Lemma 9.19 this gives an embedding into a space that does not depend on X.)

PART II

Finite-dimensional attractors

10

Partial differential equations and nonlinear semigroups

The second part of this book concentrates on the implications of Theorem 8.1 (embedding into $\mathbb{R}^k$ for sets with finite upper box-counting dimension) for the attractors of infinite-dimensional dynamical systems.

10.1 Nonlinear semigroups and attractors

We will consider (for the most part) abstract dynamical systems defined on a real Banach space $\mathscr{B}$ with the dynamics given by a nonlinear semigroup of solution operators, $S(t) : \mathscr{B} \to \mathscr{B}$ defined for $t \geq 0$, that satisfy

(i) $S(0) = \text{id}$,
(ii) $S(t)S(s) = S(t+s)$ for all $t, s \geq 0$, and
(iii) $S(t)x$ continuous in t and x.

Such semigroups can be generated by the solutions of partial differential equations, as we will outline in Sections 10.3 and 10.4. (At other points it will be useful to consider instead a dynamical system that arises from iterating a fixed function $S : \mathscr{B} \to \mathscr{B}$; such a map could be derived from a continuous time system by setting $S = S(T)$ for some fixed $T > 0$.) An attractor for $S(\cdot)$ is a compact invariant set that attracts all bounded sets.

The general theory of such semigroups and their attractors is covered in detail in Chepyzhov & Vishik (2002), Chueshov (2002), Hale (1988), Robinson (2001), Sell & You (2002), and Temam (1988); we give a brief overview of the existence theory for attractors in Chapter 11, and discuss a very general method for showing that an attractor has finite upper box-counting dimension in Chapter 12.

An attractor $\mathscr{A}$ has two properties that distinguish it from the more abstract sets we have considered in Part I: if the semigroup $S(t)$ arises from a partial differential equation then $\mathscr{A}$ consists of functions (defined on some domain U), usually with specific smoothness properties; and there are dynamics associated with the attractor.

We will show that the smoothness of functions that make up $\mathscr{A}$ can be used to obtain a bound on the thickness of $\mathscr{A}$ (Lemma 13.1), and that more dynamical properties can be used for many examples to show that the Lipschitz deviation of $\mathscr{A}$ is in fact zero (Theorem 13.3).

New questions also arise given the two properties above: it is natural to ask whether one can find an embedding consisting of point values of the functions that make up the attractor, and whether there is any 'dynamical embedding'. These questions are answered positively here in Chapters 14 and 15, under the assumption that the attractor has finite upper box-counting dimension (the result in terms of point values also requires the attractor to consist of real analytic functions).

10.2 Sobolev spaces and fractional power spaces

Since throughout Part II we will be primarily concerned with properties of semigroups generated by the solutions of partial differential equations, from time to time we will require some of the modern language used in the study of partial differential equations, in particular Sobolev spaces and a little of the theory of linear operators. We give a very cursory summary here; more detail can be found in Evans (1998) or Robinson (2001), for example.

Let $\Omega \subset \mathbb{R}^n$ be an open set with a smooth boundary. The basic space of functions upon which everything else is built is $L^2(\Omega)$, the space of (Lebesgue) square integrable functions with norm

$$\|f\|_{L^2}^2 = \int_\Omega |f(x)|^2 \, \mathrm{d}x.$$

A function f has weak derivative $D_i f = g$ (we abbreviate $\partial/\partial x_i$, the partial derivative in the ith coordinate direction, to D_i) if $g \in L^1_{\mathrm{loc}}(\Omega)$ and

$$\int_\Omega f(x)(D_i\varphi)(x) \, \mathrm{d}x = -\int_\Omega g(x)\varphi(x) \, \mathrm{d}x$$

for every $\varphi \in C_c^\infty(\Omega)$ (infinitely differentiable functions with compact support in Ω).

We use the standard notation $H^s(\Omega)$ for the Sobolev space of functions that, together with their (weak) partial derivatives of order $\leq s$, are square integrable on $\Omega \subset \mathbb{R}^n$; this space is a Hilbert space when equipped with the H^s-norm,

$$\|f\|_{H^s(\Omega)}^2 = \sum_{|\alpha| \le s} \|D^\alpha f\|_{L^2(\Omega)}^2,$$

where $\alpha = (\alpha_1, \ldots, \alpha_n)$ is a multi-index, $|\alpha| = \alpha_1 + \cdots + \alpha_n$, and

$$D^\alpha f = D_1^{\alpha_1} \cdots D_n^{\alpha_n} f = \frac{\partial^{|\alpha|}}{\partial x_1^{\alpha_1} \cdots \partial x_n^{\alpha_n}} f.$$

We will require little of the detailed theory of Sobolev spaces, but will regularly make use of the embedding result[1]

$$H^s(\Omega) \subset C^r(\Omega) \qquad \text{with} \qquad \|u\|_{C^r} \le C_{r,s} \|u\|_{H^s} \tag{10.1}$$

whenever $s > r + (n/2)$; see Exercise 10.1 for a simple proof when $n = s = 1$ and $r = 0$.

It is often useful to move between Sobolev spaces and fractional power spaces of certain linear operators (e.g. the Laplacian), since norms of fractional powers can be easier to manipulate.

Let H be a Hilbert space with norm $\|\cdot\|$ and inner product $(\cdot, \cdot)$, and let A be an unbounded positive linear operator with compact inverse that acts on H. The Hilbert–Schmidt Theorem applied to A^{-1} guarantees that A has a set of orthonormal eigenfunctions $\{w_j\}_{j=1}^\infty$ with corresponding positive eigenvalues λ_j, $Aw_j = \lambda_j w_j$, which form a basis for H (see Theorem 3.18 and Corollary 3.26 in Robinson (2001), for example), and are such that

$$Au = \sum_{j=1}^\infty \lambda_j (u, w_j) w_j \qquad \text{for all} \qquad u \in H.$$

In this setting it is straightforward to define the fractional powers of A, A^α, by

$$A^\alpha u = \sum_{j=1}^\infty \lambda_j^\alpha (u, w_j) w_j \qquad \text{for all} \qquad u \in H, \tag{10.2}$$

and if we denote by $D(A^\alpha)$ the domain in H of A^α, (i.e. $u \in H$ such that $A^\alpha u \in H$) it follows that

$$D(A^\alpha) = \left\{ \sum_{j=1}^\infty c_j w_j : \sum_{j=1}^\infty \lambda_j^{2\alpha} |c_j|^2 < \infty \right\}.$$

This space is a Hilbert space when equipped with the norm

$$\|u\|_\alpha = \|A^\alpha u\|.$$

[1] This result needs to be understood in the sense that any $f \in H^s(\Omega)$ with $s > r + (n/2)$ is equal almost everywhere to a function in $C^r(\Omega)$.

We now recall some basic properties of these fractional power spaces:

(i) If $\beta > \alpha$ then $D(A^\beta)$ is a subset of $D(A^\alpha)$, and the embedding is compact, i.e. a bounded subset of $D(A^\beta)$ is a compact subset of $D(A^\alpha)$ (see Exercise 10.2).

(ii) If A is a second order linear elliptic operator with constant coefficients then there exist constants C_s and C_s' such that

$$C_s \|A^{s/2}u\| \leq \|u\|_{H^s} \leq C_s' \|A^{s/2}u\| \qquad \text{for all} \qquad u \in D(A^{s/2}). \quad (10.3)$$

(The first of these is straightforward, the second relies on the theory of elliptic regularity; see Evans (1998), Gilbarg & Trudinger (1983), or Proposition 6.18 in Robinson (2001) for a proof when $A = -\Delta$ with Dirichlet boundary conditions.)

(iii) On a periodic domain Ω, if (a) $A = -\Delta$ along with the condition that $\int_\Omega = 0$, or if (b) $A = I - \Delta$, then $D(A^{s/2}) = H^s(\Omega)$ (see Section 6.3 in Robinson (2001)).

10.3 Abstract semilinear parabolic equations

Throughout Part II we will use two illustrative examples. The first is an abstract semilinear parabolic equation, a framework into which many particular models fit. We follow Henry (1981), but with some simplifications since we will take A to be an unbounded positive linear operator with compact inverse that acts on a Hilbert space H (as above), rather than a more general 'sectorial operator' (see Henry (1981) for details). We consider semilinear parabolic equations of the form

$$\mathrm{d}u/\mathrm{d}t = -Au + g(u) \qquad u(0) = u_0 \in D(A^\alpha), \qquad (10.4)$$

where $g(u)$ is locally Lipschitz from $D(A^\alpha)$ into H,

$$\|g(u) - g(v)\| \leq L(R)\|u - v\|_\alpha \quad \text{whenever} \quad \|u\|_\alpha, \|v\|_\alpha \leq R, \qquad (10.5)$$

for some $\alpha \in [0, 1)$. Given any $u_0 \in D(A^\alpha)$, there exists a unique solution $u(t; u_0) : [0, T) \to D(A^\alpha)$, where T depends on $\|u_0\|_\alpha$, and this solution is given by the variation of constants formula

$$u(t; u_0) = \mathrm{e}^{-At}u_0 + \int_0^t \mathrm{e}^{-A(t-s)}g(u(s))\,\mathrm{d}s, \qquad (10.6)$$

where

$$e^{-At}u = \sum_{j=1}^{\infty} e^{-\lambda_j t}(u, w_j)w_j$$

(see Henry (1981, Lemma 3.3.2 and Theorem 3.3.3)). Solutions are continuous from $[0, T)$ into $D(A^\alpha)$ and depend continuously on the initial condition (Henry's Theorem 3.4.1).

If we assume that unique solutions of (10.4) exist for all $t \geq 0$ (this usually requires an equation-by-equation approach tailored to the particular model under consideration), then the solutions generate a semigroup on $D(A^\alpha)$ via the definition $S(t)u_0 = u(t; u_0)$. Properties (i) and (iii) are immediate from the above results, and property (ii), $S(t+s) = S(t)(s)$ for all $t, s \geq 0$, follows from the uniqueness of solutions.

The following estimates for the action of e^{-At} between different fractional power spaces are extremely useful:

$$\|e^{-At}\|_{\mathscr{L}(H,D(A^\gamma))} = \|A^\gamma e^{-At}\|_{\mathscr{L}(H)} \leq \begin{cases} \gamma^\gamma e^{-\gamma} t^{-\gamma} & 0 < t < \gamma/\lambda_1, \\ \lambda_1^\gamma e^{-\lambda_1 t} & t \geq \gamma/\lambda_1, \end{cases} \tag{10.7}$$

and consequently

$$\int_0^\infty \|A^\gamma e^{-At}\|_{\mathscr{L}(H)}\, dt \leq I_\gamma := \frac{e^{-\gamma}\lambda_1^{-(1-\gamma)}}{1-\gamma} \quad \text{if } \gamma \in [0, 1), \tag{10.8}$$

see Exercise 10.3. Sometimes it is convenient to rewrite (10.7) as

$$\|A^\gamma e^{-At}\|_{\mathscr{L}(H)} \leq c_\delta t^{-\gamma} e^{-\delta t}, \tag{10.9}$$

where $0 < \delta < \lambda_1$ and

$$c_\delta = \gamma^\gamma e^{-\gamma} \max(e^{\delta\gamma/\lambda_1}, (\lambda_1 - \delta)^{-\gamma}). \tag{10.10}$$

10.4 The two-dimensional Navier–Stokes equations

Our other example will be the two-dimensional Navier–Stokes equations. The existence of unique solutions for all $t \geq 0$ in the three-dimensional case is a well-known unsolved problem (and is one of the Clay Foundation's Million Dollar Millennium Prize Problems); the dynamical systems theory has therefore concentrated mainly on the two-dimensional case, for which suitable existence and uniqueness results are available. We will do the same here.

Classically, these are equations for the two-component velocity $u(x,t)$ and the scalar pressure $p(t)$,

$$\frac{\partial u}{\partial t} - \Delta u + (u \cdot \nabla)u + \nabla p = f(x) \qquad \nabla \cdot u = 0, \tag{10.11}$$

where we have set the kinematic viscosity (the coefficient of the Laplacian term) equal to 1. The right-hand side f represents a body forcing that maintains the motion (with $f = 0$ every solution decays to zero and the attractor is trivial, see Exercise 10.5).

For mathematical simplicity we will concentrate on the periodic case, when $x \in \Omega = [0, 2\pi]^2$ and

$$u(x + 2\pi e_i, t) = u(x,t), \qquad i = 1, 2, \tag{10.12}$$

where e_1 and e_2 are orthonormal vectors in $\mathbb{R}^2$. It is also convenient to assume the zero-average conditions

$$\int_\Omega f(x)\,\mathrm{d}x = 0 \qquad \text{and} \qquad \int_\Omega u_0(x)\,\mathrm{d}x = 0; \tag{10.13}$$

the condition on f ensures that the zero average of $u(x,t)$ is preserved under the time evolution.

The natural phase space for the problem we will denote[2] by H: it is the completion in the $[L^2(\Omega)]^2$-norm of

$$\mathscr{H} = \{u \in [C^\infty_{\text{per}}(\Omega)]^2 : \nabla \cdot u = 0 \quad \text{and} \quad \int_\Omega u(x)\,\mathrm{d}x = 0\}, \tag{10.14}$$

where $C^\infty_{\text{per}}(\Omega)$ is the space of all C^∞ functions that are periodic as in (10.12); we equip H with the L^2 norm. (Roughly speaking H consists of all functions in $[L^2(\Omega)]^2$ with zero average and (generalised) divergence zero.)

Given $f \in H$ and $u_0 \in H$, for all $t \geq 0$ there exists a unique solution which we denote by $u(t) = u(t; u_0)$ (suppressing the x dependence) that is continuous from $[0, \infty)$ into H, and depends continuously (in the H-norm) on u_0 (see Constantin & Foias (1988), Robinson (2001), or Temam (1977)). As with the abstract semilinear equation in the previous section, we can use the solution to define a semigroup on H by setting $S(t)u_0 = u(t)$.

It is a standard approach (particularly in the literature that views the two-dimensional equations as a dynamical system) to reformulate (10.11) in 'functional form', essentially as an ordinary differential equation on an appropriate space. This reformulation is one way to eliminate the pressure from the equations, capitalising on the observation that the pressure term ∇p is orthogonal

[2] Whenever we are dealing with the Navier–Stokes equations our 'primary' Hilbert space will be H, so this should not cause any confusion with more general abstract considerations.

(in L^2) to functions that are divergence free ($\nabla \cdot u = 0$):

$$\int_\Omega u \cdot \nabla p \,\mathrm{d}x = -\int_\Omega p(\nabla \cdot u)\,\mathrm{d}x = 0. \tag{10.15}$$

To recast the equation in this form, let V denote the completion of $\mathscr{H}$ in the $[H^1(\Omega)]^2$ norm, and let V^* denote the dual of V. Taking the inner product of (10.11) with $v \in V$, after some integrations by parts (in particular using (10.15)) we obtain[3]

$$\frac{\mathrm{d}}{\mathrm{d}t}(u, v) + (Du, Dv) + ((u \cdot \nabla)u, v) = (f, v) \qquad \text{for all} \qquad v \in V.$$

Now define a linear operator $A : V \to V^*$ by

$$\langle Au, v\rangle = (Du, Dv) \qquad \text{for all} \qquad u, v \in V,$$

where $\langle \cdot, \cdot\rangle$ denotes the pairing between V^* and V, and define a bilinear form $B : V \times V \to V^*$ by

$$\langle B(u, u), v\rangle = ((u \cdot \nabla)u, v) \qquad \text{for all} \qquad u, v \in V.$$

Then we can rewrite the Navier–Stokes equations as an ordinary differential equation in the space V^*:

$$\frac{\mathrm{d}u}{\mathrm{d}t} + Au + B(u, u) = f.$$

The operator A is the 'Stokes operator', and is given by $A = -\Pi\Delta$, where Π is the orthogonal projection in $[L^2(\Omega)]^2$ onto H (divergence-free vector fields). In the periodic case, $A = -\Delta$ on its domain of definition, and so (see property (iii) of fractional power spaces) the Sobolev spaces $H^s(\Omega)$ coincide with the fractional power spaces $D(A^{s/2})$ with

$$c_s\|A^{s/2}u\| \le \|u\|_{H^s} \le C_s\|A^{s/2}u\|$$

for some c_s, C_s.

In the analysis that follows we will only require the following properties of A and B:

$$(Au, u) = \|Du\|^2 \qquad \text{for all} \qquad u \in V, \tag{10.16}$$

which follows immediately from the definition of A, and two orthogonality relations for the nonlinear term,

$$(B(u, v), v) = 0 \qquad \text{for all} \qquad u, v \in V \tag{10.17}$$

[3] The notation (Du, Dv) denotes $\sum_{i,j=1}^2 (D_i u_j, D_i v_j)$; in particular $\|Du\|^2 = \sum_{i,j=1}^2 \|D_i u_j\|^2$.

(this follows from an integration by parts, and remains true for other boundary conditions and in the three-dimensional case), and

$$(B(u,u), Au) = 0 \qquad \text{for all} \qquad u \in D(A), \tag{10.18}$$

which is only true in the two-dimensional periodic case (the proof relies on expanding the expression $((u \cdot \nabla)u, \Delta u)$, then using the divergence-free condition repeatedly in many pairwise cancellations). We will also make use of the Poincaré inequality,

$$\|u\| \le \|Du\| \qquad \text{for all} \qquad u \in V, \tag{10.19}$$

which follows making use of the zero-average condition (10.13), and is easy to see using the Fourier expansion of u (see Exercise 10.6).

Finally we note that the Navier–Stokes equations can be recast in the abstract form (10.4), where A is the Stokes operator and

$$g(u) = f - B(u,u).$$

In this case $g(u)$ is locally Lipschitz from $D(A^\alpha)$ into H for any $\alpha > 1/2$:

$$\begin{aligned}\|g(u) - g(v)\|_{L^2} &= \|B(u,u) - B(v,v)\|_{L^2}\\ &= \|B(u,u-v) + B(u-v,v)\|_{L^2}\\ &\le \|B(u,u-v)\|_{L^2} + \|B(u-v,v)\|_{L^2}.\end{aligned}$$

For the first term we have (since $\|u\|_\infty \le c\|u\|_{H^{2\alpha}}$ as $\alpha > 1/2$)

$$\begin{aligned}\|B(u,u-v)\|_{L^2} &\le \|u\|_\infty \|D(u-v)\|_{L^2}\\ &\le c\|u\|_{H^{2\alpha}}\|u-v\|_{H^1}\\ &\le c\|u\|_\alpha \|u-v\|_{1/2},\end{aligned}$$

while for the second term we use Hölder's inequality to write

$$\|B(u-v,v)\|_{L^2} \le \|u-v\|_{L^{2/(2\alpha-1)}}\|Dv\|_{L^{1/(1-\alpha)}}.$$

For the first and final time we use the two-dimensional Sobolev embedding result (see Evans (1998), for example)

$$H^s \subset L^{2/(1-s)} \qquad \text{with} \qquad \|u\|_{L^{2/(1-s)}} \le c_s\|u\|_{H^s}$$

(which in particular shows that any L^p norm is bounded by the H^1 norm) with $s = 2\alpha - 1$ and obtain

$$\|B(u-v,v)\| \le c\|u-v\|_{H^1}\|v\|_{H^{2\alpha}} \le c\|u-v\|_{1/2}\|v\|_\alpha.$$

It follows that

$$\|g(u) - g(v)\|_{L^2} \le c[\|u\|_\alpha + \|v\|_\alpha]\|u - v\|_{1/2}, \tag{10.20}$$

and in general g is locally Lipschitz from $D(A^\alpha)$ into H.

One could therefore could treat the two-dimensional Navier–Stokes equations within the abstract framework of Section 10.3. However, this would require us to take an initial condition in $D(A^\alpha)$ with $\alpha > 1/2$, and to consider the dynamical system generated on this space. Instead, it is more useful to obtain the existence of a solution via other methods (above we stated that a unique solution exists for any $u_0 \in H$), and then use the more abstract setting when it makes the analysis more convenient.

We will adopt this approach in Section 13.2, in which we investigate the Lipschitz deviation of attractors. There, the following observation will be central: it follows from (10.20) that if we restrict our attention to a set X that is bounded in $D(A^\alpha)$ with $\alpha > \frac{1}{2}$ then g is Lipschitz from $D(A^{1/2})$ into H:

$$\|g(u) - g(v)\|_{L^2} \le C\|u - v\|_{1/2} \qquad \text{for all} \qquad u, v \in X.$$

Exercises

10.1 If $f(x) = \sum_{k\in\mathbb{Z}} c_k e^{ikx}$ then $f \in H^1(0, 2\pi)$ provided that

$$\|f\|_{H^1}^2 = \sum (1 + |k|^2)|c_k|^2 < \infty.$$

Show that $\|f\|_\infty \le c\|f\|_{H^1}$, and deduce that $f \in C^0([0, 2\pi])$.

10.2 Show that $D(A^\beta)$ is compactly embedded in if $f \epsilon H^1|0, 2\pi)$ then $D(A^\alpha)$ if $\beta > \alpha$.

10.3 If $u = \sum_{j=1}^\infty c_j w_j$ then

$$\|A^\gamma e^{-At} u\|^2 = \sum_{j=1}^\infty \lambda_j^{2\gamma} e^{-2\lambda_j t} |c_j|^2.$$

Derive the bounds in (10.7) and (10.8).

10.4 Suppose that a solution $u(t) \in D(A^\alpha)$ of (10.4) exists on $[0, T)$. Use the variation of constants formula (10.6) and the estimates in (10.7) and (10.9) to show that $u(t) \in D(A^\beta)$ for all $\beta < 1$.

10.5 Consider the two-dimensional Navier–Stokes equations in functional form with zero forcing:

$$du/dt + Au + B(u, u) = 0.$$

Show that $\|u(t)\|_{L^2} \to 0$ as $t \to \infty$.

10.6 For $u \in V$ (recall that V is the completion of $\mathscr{H}$, defined in (10.14), in the $[H^1(\Omega)]^2$ norm) prove the Poincaré inequality $\|u\| \leq \|Du\|$ using the Fourier expansion

$$u = \sum_{k \in \dot{\mathbb{Z}}^2} c_k \mathrm{e}^{\mathrm{i}k \cdot x},$$

where $\dot{\mathbb{Z}}^2 = \mathbb{Z}^2 \setminus \{0, 0\}$ (there is no $k = 0$ term since $\int_\Omega u = 0$ for $u \in V$).

11
Attracting sets in infinite-dimensional systems

11.1 Global attractors

In this chapter we give a basic existence result for attractors (Theorem 11.3), and show in Proposition 11.4 that the attractor can be characterised in an 'analytical' way that is independent of the dynamical definition that is the primary one here. Sections 11.3 and 11.4 prove the existence of attractors for the models introduced in the previous chapter.

A set $X \subset \mathscr{B}$ is said to be *invariant* if $S(t)X = X$ for all $t \geq 0$, and is said to attract $B \subset \mathscr{B}$ if

$$\operatorname{dist}(S(t)B, X) \to 0 \qquad \text{as} \qquad t \to \infty.$$

A set $X \subset \mathscr{B}$ is said to be *attracting* if it attracts all bounded subsets of $\mathscr{B}$.

A set $\mathscr{A} \subset \mathscr{B}$ is said to be *the global attractor* if it is compact, invariant, and attracting. If it exists then the global attractor is unique: suppose that $\mathscr{A}_1$ and $\mathscr{A}_2$ are two global attractors. Then, since $\mathscr{A}_2$ is bounded, it is attracted by $\mathscr{A}_1$,

$$\operatorname{dist}(S(t)\mathscr{A}_2, \mathscr{A}_1) \to 0 \qquad \text{as} \qquad t \to \infty.$$

But $\mathscr{A}_2$ is invariant, $S(t)\mathscr{A}_2 = \mathscr{A}_2$, and so $\operatorname{dist}(\mathscr{A}_2, \mathscr{A}_1) = 0$. The argument is symmetric, so $\operatorname{dist}(\mathscr{A}_1, \mathscr{A}_2) = 0$, from which it follows that $\mathscr{A}_1 = \mathscr{A}_2$.

Two alternative characterisations of the attractor follow from a similar argument: $\mathscr{A}$ is the maximal compact invariant set, and the minimal closed set that attracts all bounded sets, see Exercise 11.1.

11.2 Existence of the global attractor

In this section we will use the following simple lemma repeatedly.

Lemma 11.1 *Let K be a compact subset of $\mathscr{B}$, and $x_n \in \mathscr{B}$ a sequence with $\lim_{n\to\infty} \operatorname{dist}(x_n, K) = 0$. Then $\{x_n\}$ has a convergent subsequence, whose limit lies in K.*

Proof Write $x_n = k_n + z_n$, where $k_n \in K$ and $\|z_n\| \to 0$ as $n \to \infty$. Then there is a subsequence such that $k_{n_j} \to k^* \in K$, so $x_{n_j} \to k^*$ too. □

We will in fact form the attractor as the union of the omega-limit sets (defined in the following lemma) of all possible bounded sets. We begin by proving some properties of these limit sets.

Proposition 11.2 *Suppose that there exists a compact attracting set K. Then for any bounded set B, the set*

$$\omega(B) = \bigcap_{t\geq 0} \overline{\bigcup_{s\geq t} S(s)B} \tag{11.1}$$

$$= \{x \in \mathscr{B} : x = \lim_{n\to\infty} S(t_n)b_n \text{ for some } t_n \to \infty,\ b_n \in B\} \tag{11.2}$$

is a nonempty compact subset of K that is invariant and attracts B.

For the equivalence of (11.1) and (11.2) see Exercise 11.2.

Proof Since there is a compact attracting set, Lemma 11.1 combined with (11.2) shows that $\omega(B) \subseteq K$; that $\omega(B)$ is nonempty follows similarly, taking any initial sequences $\{b_n\} \in B$ and $t_n \to \infty$. Using (11.1), $\omega(B)$ is a decreasing sequence of closed sets, and so is a closed subset of the compact set K; thus $\omega(B)$ is compact.

Now suppose that $x \in \omega(B)$. Then there exist sequences $\{t_n\}$ with $t_n \to \infty$ and $\{b_n\}$ with $b_n \in B$ such that $x = \lim_{n\to\infty} S(t_n)b_n$. Then, since $S(t)$ is continuous,

$$S(t)x = S(t)\left(\lim_{n\to\infty} S(t_n)b_n\right) = \lim_{n\to\infty} S(t + t_n)b_n,$$

and so $S(t)x \in \omega(B)$, i.e. $S(t)\omega(B) \subseteq \omega(B)$.

Now, if $y \in \omega(B)$ then $y = \lim_{n\to\infty} S(t_n)b_n$. For any fixed t, once $t_n \geq t$, we can write $S(t_n)b_n = S(t)[S(t_n - t)b_n]$. Using Lemma 11.1, we know that $S(t_n - t)b_n$ has a convergent subsequence, which converges to some $\beta \in \omega(B)$. Taking the limit through this subsequence, it follows that $y = S(t)\beta$ with $\beta \in \omega(B)$, so $\omega(B) \subseteq S(t)\omega(B)$, and hence $S(t)\omega(B) = \omega(B)$.

We now show that $\omega(B)$ attracts B. If not, then there exist a $\delta > 0$, and $t_n \to \infty$, $b_n \in B$ such that

$$\operatorname{dist}(S(t_n)b_n, \omega(B)) > \delta.$$

But (by Lemma 11.1) $\{S(t_n)b_n\}$ has a convergent subsequence, whose limit must lie in $\omega(B)$, a contradiction. □

The existence theorem for global attractors is essentially an immediate corollary of the above.

Theorem 11.3 *There exists a global attractor $\mathscr{A}$ if and only if there exists a compact attracting set K, in which case $\mathscr{A} = \omega(K)$.*

Proof If $\mathscr{A}$ is an attractor then it is a compact attracting set; since $S(t)\mathscr{A} = \mathscr{A}$ for all $t \geq 0$ we have $\omega(\mathscr{A}) = \mathscr{A}$. Conversely, if K is a compact attracting set then Proposition 11.2 shows that for every bounded set B, the omega-limit set $\omega(B)$ is compact, invariant, and attracts B. Define

$$\mathscr{A} = \overline{\bigcup_{B \text{ bounded}} \omega(B)}. \tag{11.3}$$

The set $\mathscr{A}$ is clearly compact (since each $\omega(B)$ is contained in the compact set K), invariant, and attracts every bounded set B, so is the global attractor. It only remains to show that $\mathscr{A} = \omega(K)$. It is immediate from (11.3) that $\mathscr{A} \supseteq \omega(K)$, while since $\mathscr{A}$ is the minimal closed set that attracts bounded sets (Exercise 11.1) we must have $\mathscr{A} \subseteq K$, and hence $\mathscr{A} = \omega(\mathscr{A}) \subseteq \omega(K)$. □

In many cases we can show something stronger than the existence of a compact attracting set, namely the existence of a compact absorbing set. We say that a set $X \subset \mathscr{B}$ is *absorbing* if for every bounded subset $B \subset \mathscr{B}$ there exists a time t_B such that

$$S(t)B \subseteq X \qquad \text{for all} \qquad t \geq t_B,$$

i.e. the orbits of all bounded sets eventually enter and do not leave X. Clearly the existence of a compact absorbing set implies the existence of a compact attracting set, which we know implies the existence of a global attractor. In Section 11.3 we prove that the existence of a bounded absorbing set for an abstract semilinear parabolic equation implies the existence of a compact absorbing set, and hence of the global attractor; in Section 11.4 we will prove the existence of a global attractor for the two-dimensional Navier–Stokes equations by showing directly the existence of a compact absorbing set.

Finally we give an alternative, more analytical characterisation of attractors in terms of complete bounded orbits. This shows that while these objects have a definition in terms of dynamics, they are of interest independent of their dynamical interpretation.

Proposition 11.4 *The global attractor is given by*

$$\mathscr{A} = \{u_0 \in \mathscr{B} : \text{ there exists a solution } u(t) \text{ defined for all } t \in \mathbb{R} \text{ with } u(0) = u_0 \text{ such that } \|u(t)\| \le M \ \forall\, t \in \mathbb{R}, \text{ for some } M > 0\}.$$

Proof Suppose that there is a globally bounded solution through u_0, i.e. that u_0 is an element of the set on the right-hand side of the identity in the proposition. Then for every $t \ge 0$, $u_0 = S(t)u(-t)$ with $\|u(-t)\| \le M$, i.e. $u(-t) \in B(0, M)$. Then

$$\mathrm{dist}(u_0, \mathscr{A}) = \mathrm{dist}(S(t)u(-t), \mathscr{A}) \le \mathrm{dist}(S(t)B(0, M), \mathscr{A}) \to 0 \quad \text{as } t \to \infty,$$

and so $u_0 \in \mathscr{A}$. To prove the opposite inclusion, given $u_0 \in \mathscr{A}$ it is clear that $\|S(t)u_0\|$ is bounded for all $t \ge 0$, since $S(t)u_0 \in \mathscr{A}$. To extend the solution backwards in time, first find $u(-1) \in \mathscr{A}$ such that $S(1)u(-1) = u_0$ (this is possible since $S(1)\mathscr{A} = \mathscr{A}$), and let

$$u(-1 + t) = S(t)u(-1) \qquad \text{for} \qquad t \in [0, 1).$$

Continue inductively: choose $u(-(n+1))$ such that $u(-n) = S(1)u(-(n+1))$ and define

$$u(-(n+1) + t) = S(t)u(-(n+1)) \qquad \text{for} \qquad t \in [0, 1).$$

The semigroup property ensures that this gives a solution, and since the solution lies within $\mathscr{A}$ for all $t \in \mathbb{R}$ it is bounded. □

11.3 Example 1: semilinear parabolic equations

First we consider the semilinear parabolic equation of the form

$$\mathrm{d}u/\mathrm{d}t = -Au + g(u) \qquad u(0) = u_0 \in D(A^\alpha), \tag{11.4}$$

where g is locally Lipschitz from $D(A^\alpha)$ into H,

$$\|g(u) - g(v)\| \le L(R)\|u - v\|_\alpha \quad \text{whenever} \quad \|u\|_\alpha, \|v\|_\alpha \le R, \tag{11.5}$$

as introduced above in Section 10.3. Recall that the solution $u(t)$ of (11.4) is given by the variation of constants formula

$$u(t; u_0) = \mathrm{e}^{-At}u_0 + \int_0^t \mathrm{e}^{-A(t-s)}g(u(s))\,\mathrm{d}s.$$

Assume that there is a bounded absorbing set in $D(A^\alpha)$: i.e. that for any R_0 there exists a $t_0(R_0)$ such that

$$\|u(t)\|_\alpha \le M \qquad \text{for all} \qquad t \ge t_0(R_0)$$

(for particular equations such an estimate is usually proved using model-specific techniques, rather than appealing to a general result for (11.4)). Given this bounded absorbing set in $D(A^\alpha)$, one can use the variation of constants formula to obtain a compact absorbing set, as we now show.

Note that the Lipschitz property of g in (11.5) implies that

$$|g(u)| \le M' := |f(0)| + L(M)M \qquad \text{when} \quad \|u\|_\alpha \le M.$$

Therefore, choosing $\epsilon > 0$ such that $\alpha + \epsilon < 1$,

$$\begin{aligned}\|u(t)\|_{\alpha+\epsilon} &= \|A^{\alpha+\epsilon}u(t)\| \\ &= \left\| A^{\alpha+\epsilon}\mathrm{e}^{-A(t-t_0)}u(t_0) + \int_{t_0}^t A^{\alpha+\epsilon}\mathrm{e}^{-A(t-s)}g(u(s))\,\mathrm{d}s \right\| \\ &\le \|A^\epsilon \mathrm{e}^{-A(t-t_0)}\|_{\mathscr{L}(H)}\|u(t_0)\|_\alpha + \int_{t_0}^t \|A^{\alpha+\epsilon}\mathrm{e}^{-A(t-s)}\|_{\mathscr{L}(H)}M'\,\mathrm{d}s,\end{aligned}$$

and using the estimates

$$\|A^\gamma \mathrm{e}^{-At}\|_{\mathscr{L}(H)} \le ct^{-\gamma}\mathrm{e}^{-\delta t} \qquad \text{and} \qquad \int_0^\infty \|A^\gamma \mathrm{e}^{-As}\|_{\mathscr{L}(H)}\,\mathrm{d}s \le I_\gamma < \infty$$

from (10.9) and (10.8), one can obtain

$$\|u(t)\|_{\alpha+\epsilon} \le c(t-t_0)^{-\epsilon}\mathrm{e}^{-\delta(t-t_0)}\|u_0\|_\alpha + I_{\alpha+\epsilon}M'.$$

In particular, for all $t \ge t_0(R_0) + 1$,

$$\|u(t)\|_{\alpha+\epsilon} \le cM + I_{\alpha+\epsilon}M',$$

and so there is a bounded absorbing set in $D(A^{\alpha+\epsilon})$. Since $D(A^{\alpha+\epsilon})$ is compactly embedded in $D(A^\alpha)$ (see property (i) at the end of Section 10.2), there is a compact absorbing set in $D(A^\alpha)$, and hence a global attractor for the semigroup defined on $D(A^\alpha)$.

11.4 Example 2: the two-dimensional Navier–Stokes equations

We consider the two-dimensional Navier–Stokes equations written in their functional form

$$\frac{\mathrm{d}u}{\mathrm{d}t} + Au + B(u,u) = f, \qquad f \in H, \tag{11.6}$$

see Section 10.4, and prove the existence of a compact absorbing set in H, by showing the existence of a bounded absorbing set in V (recall that V is compactly embedded in H).

First, if one takes the inner product (in H) of (11.6) with u, since $(Au, u) = \|Du\|^2$ (10.16) and $(B(u, u), u) = 0$ (10.17) one obtains

$$\frac{1}{2}\frac{\mathrm{d}}{\mathrm{d}t}\|u\|^2 + \|Du\|^2 = (f, u) \leq \|f\|\|u\|. \tag{11.7}$$

Using the Poincaré inequality $\|u\| \leq \|Du\|$ (10.19) on the left-hand side, and Young's inequality ($2ab \leq a^2 + b^2$) on the right-hand side, one obtains the differential inequality

$$\frac{\mathrm{d}}{\mathrm{d}t}\|u\|^2 + \|u\|^2 \leq \|f\|^2.$$

This can be readily integrated (using the integrating factor e^t) to deduce that

$$\|u(t)\|^2 \leq \|u_0\|^2\mathrm{e}^{-t} + \|f\|^2(1 - \mathrm{e}^{-t}),$$

and so $\|u(t)\|^2 \leq 2\|f\|^2$ for all $t \geq t_0(\|u_0\|)$. This provides a *bounded* absorbing set in H. To obtain a compact absorbing set, we will show that there is a bounded absorbing set in H^1, i.e. that $\|Du(t)\|^2$ is asymptotically bounded, uniformly in terms of the L^2 norm of the initial condition.

We first require a subsidiary estimate. Dealing with (11.7) differently, one can use the Poincaré inequality and Young's inequality on the right-hand side to obtain

$$\frac{\mathrm{d}}{\mathrm{d}t}\|u\|^2 + \|Du\|^2 \leq \|f\|^2.$$

Integrating this differential inequality from t to $t + 1$ gives

$$\|u(t + 1)\|^2 + \int_t^{t+1} \|Du(s)\|^2\,\mathrm{d}s \leq \|f\|^2 + \|u(t)\|^2.$$

In particular, therefore, since $\|u(t)\|^2 \leq 2\|f\|^2$ for $t \geq t_0(\|u_0\|)$,

$$\int_t^{t+1} \|Du(s)\|^2\,\mathrm{d}s \leq 3\|f\|^2 \qquad \text{for all} \qquad t \geq t_0(\|u_0\|). \tag{11.8}$$

Now take the inner product of (11.6) with Au, and use the special two-dimensional periodic orthogonality relation (10.18) to obtain

$$\frac{1}{2}\frac{\mathrm{d}}{\mathrm{d}t}\|Du\|^2 + \|Au\|^2 = (f, Au) \leq \|f\|\|Au\| \leq \frac{1}{2}\|f\|^2 + \frac{1}{2}\|Au\|^2. \tag{11.9}$$

Absorbing the $\|Au\|^2$ from the right-hand side into the same term on the left-hand side, and dropping the resulting $+\frac{1}{2}\|Au\|^2$, we obtain

$$\frac{\mathrm{d}}{\mathrm{d}t}\|Du\|^2 \leq \|f\|^2. \tag{11.10}$$

Take $t \geq t_0(\|u_0\|)$, and integrate (11.10) from s to $t+1$, where $t \leq s \leq t+1$:

$$\|Du(t+1)\|^2 \leq \|f\|^2 + \|Du(s)\|^2.$$

Now integrate once more, but this time with respect to s between t and $t+1$, which yields

$$\|Du(t+1)\|^2 \leq \|f\|^2 + \int_t^{t+1} \|Du(s)\|^2 \leq 4\|f\|^2,$$

using (11.8). Since this is valid for all $t \geq t_0(\|u_0\|)$, it follows that

$$\|Du(t)\|^2 \leq 6\|f\|^2 \qquad \text{for all} \qquad t \geq t_0(\|u_0\|) + 1.$$

(This 'double integration' trick can be formalised as the 'Uniform Gronwall Lemma', see Exercise 11.5.)

This implies the existence of a bounded absorbing set in H^1, and since H^1 is compactly embedded in L^2, this gives a compact absorbing set in H and guarantees the existence of a global attractor $\mathscr{A}$ for the semigroup on H. With a little further work one can show that the global attractor is a bounded subset of $D(A)$ (and hence of H^2), see Exercise 11.6.

Exercises

11.1 Show that the global attractor is the maximal compact invariant set, and the minimal closed set that attracts all bounded sets.

11.2 Show that (11.1) and (11.2) are equivalent.

11.3 Show that $\mathscr{A}$ is connected whenever $\mathscr{B}$ is connected. [Hint: argue by contradiction.]

11.4 If X is an invariant set, the unstable set of X is defined by

$$U(X) := \{u_0 \in \mathscr{B} : \text{ there exists a globally defined solution } u(t) \text{ with } u(0) = u_0 \text{ and } \operatorname{dist}(u(t), X) \to 0 \text{ as } t \to -\infty\}.$$

Show that $U(X) \subset \mathscr{A}$ for any invariant set X.

11.5 Use the double integration method used in Section 11.4 to prove the 'Uniform Gronwall Lemma': if x, a, and b are positive functions such that

$$\mathrm{d}x/\mathrm{d}t \leq ax + b$$

with

$$\int_t^{t+r} x(s)\,\mathrm{d}s \leq X, \qquad \int_t^{t+r} a(s), \mathrm{d}s \leq A, \qquad \text{and} \qquad \int_t^{t+r} b(s)\,\mathrm{d}s \leq B$$

for some $r > 0$ and all $t \geq t_0$, then

$$x(t) \leq \left(\frac{X}{r} + B\right) \mathrm{e}^A$$

for all $t \geq t_0 + r$. [Hint: use the integrating factor $\exp(-\int_s^t a(\tau)\,\mathrm{d}\tau)$ with $t_0 \leq t \leq s \leq t + r$.]

11.6 This exercise provides a proof that if $f \in H$ then the attractor for the two-dimensional Navier–Stokes equations is a bounded subset of $D(A)$. Fix $u_0 \in \mathscr{A}$ and let $u(t) = S(t)u_0$ be the solution with initial condition u_0. It follows from the invariance of $\mathscr{A}$ that $u(t) \in \mathscr{A}$ for all $t \geq 0$, and hence that $\|u(t)\|^2 \leq 2\|f\|^2$ and $\|Du(t)\|^2 \leq 6\|f\|^2$ for all $t \geq 0$.

(i) Starting from (11.9) show that

$$\int_0^1 \|Au(s)\|^2\,\mathrm{d}s \leq 7\|f\|^2 \qquad \text{for all} \qquad t \geq 0.$$

(ii) Given the inequality

$$\|B(u,u)\| \leq c_1\|u\|^{1/2}\|Du\|\,\|Au\|^{1/2},$$

use (11.6) to deduce that

$$\int_0^1 \|u_t(s)\|^2\,\mathrm{d}s \leq I_t \qquad \text{for all} \qquad t \geq 0,$$

where $u_t = \mathrm{d}u/\mathrm{d}t$ and I_t depends only on c_1 and $\|f\|$.

(iii) Differentiate (11.6) with respect to t, and then use the estimate

$$|(B(u,v),u)| \leq c_2\|u\|\,\|Du\|\,\|Dv\| \tag{11.11}$$

along with the uniform Gronwall (double integration) 'trick' to show that

$$\|u_t(1)\| \leq \rho_t$$

(where ρ_t depends only on c_1, c_2, and $\|f\|$).

(iv) Deduce via (11.6) that $\|Au(1)\| \leq \rho_A$ (depending only on c_1, c_2, and $\|f\|$) and use the invariance of $\mathscr{A}$ to conclude that $\mathscr{A}$ is bounded in $D(A)$.

(This method is essentially due to Heywood & Rannacher (1982).)

12
Bounding the box-counting dimension of attractors

Powerful techniques are available for bounding the box-counting dimension of attractors in Hilbert spaces, the case most often encountered in applications. The most widely-used method was developed for finite-dimensional dynamical systems by Douady & Oesterlé (1980), and was extended to treat subsets of infinite-dimensional Hilbert spaces by Constantin & Foias (1985). Much effort has also been expended in refining the resulting estimates for particular models, in particular for the two-dimensional Navier–Stokes equations (for a nice overview see Doering & Gibbon (1995)).

However, general results providing bounds on the dimension of compact invariant sets go back to Mallet-Paret (1976), who showed that if K is a compact subset of a Hilbert space H, $f : H \to H$ is continuously differentiable, $f(K) \supseteq K$ ('K is negatively invariant'), and the derivative of f is everywhere equal to the sum of a compact map and a contraction, then the upper box-counting dimension of K is finite. Mañé (1981) generalised this argument to treat subsets of Banach spaces (this was in the same paper in which he proved a 'generic' embedding theorem for sets with $d_{\rm H}(X - X)$ finite, cf. our Theorem 6.2).

The Hilbert space method is already cleanly and clearly presented in a number of texts that concentrate more specifically on estimating the dimension of attractors (e.g. Chepyzhov & Vishik, 2002; Robinson, 2001; Temam, 1988), and a general technique that covers the Banach space case seems more in keeping with the rest of this book. We therefore give here a simplified proof of Mañé's result, due to Carvalho *et al.* (2010); the Hilbert space theory is covered in Exercises 12.4–12.8.

Throughout this chapter we will use the notation $B_Z(0, r)$ to denote the ball of radius r, centred at zero, in the space Z; we will continue to use the simpler notation $B(0, r)$ for the r-ball in the Banach space $\mathscr{B}$.

All of the arguments that provide bounds on the dimension of attractors follow similar lines, which we first sketch and then make formal in a lemma. For the majority of this chapter we treat a compact set K that is invariant under the action of a map $f : \mathscr{B} \to \mathscr{B}$; f could be derived from a semigroup $S(\cdot)$ by setting $f = S(T)$ for some suitable T.

Suppose that K can be covered by N_0 balls of radius ϵ, $\{B(x_j, \epsilon)\}_{j=1}^{N_0}$. Then since $f(K) = K$, it follows that K can be covered by the images $\{f(B(x_j, \epsilon))\}_{j=1}^{N_0}$. If ϵ is sufficiently small,

$$f(B(x_j, \epsilon)) \simeq f(x_j) + \epsilon \mathrm{D}f(x_j)[B(0, 1)],$$

where $\mathrm{D}f(x_j)$ is the derivative of f at x_j. If we can find an efficient covering of $\mathrm{D}f[B(0, 1)]$ by balls of a smaller radius $\alpha < 1$, say

$$N(\mathrm{D}f(x)[B(0, 1)], \alpha) \le M \qquad \text{for all} \qquad x \in K,$$

then we have a new cover of K by MN balls of radius $\alpha\epsilon$. Iterating this procedure will give a cover of K by $M^k N$ balls of radius $\alpha^k \epsilon$, from which the bound $d_{\mathrm{B}}(K) \le \log M/(-\log \alpha)$ follows.

We now make this precise.

Lemma 12.1 *Let K be a compact subset of a Banach space $\mathscr{B}$ that is invariant for the map $f : \mathscr{B} \to \mathscr{B}$, i.e. $f(K) = K$. Suppose in addition that f is continuously differentiable on a neighbourhood of K, and that there exist α, $0 < \alpha < 1$, and $M \ge 1$ such that for any $x \in K$,*

$$N(\mathrm{D}f(x)[B(0, 1)], \alpha) \le M. \tag{12.1}$$

Then[1]

$$d_{\mathrm{B}}(K) \le \frac{\log M}{-\log \alpha}. \tag{12.2}$$

Proof First, we ensure that (12.1) is sufficient to provide a bound on the number of balls required to cover $f(B(x, r))$ when r is small enough. Since f is continuously differentiable and K is compact, for any $\eta > 0$ there exists an $r_0 = r_0(\eta)$ such that for any $0 < r < r_0$ and any $x \in K$,

$$f(B(x, r)) \subseteq f(x) + \mathrm{D}f(x)[B(0, r)] + B(0, \eta r),$$

where $A + B$ is used to denote the set $\{a + b : a \in A, b \in B\}$. It follows that

$$N(f(B(x, r)), (\alpha + \eta)r) \le M \tag{12.3}$$

for all $r \le r_0(\eta)$.

[1] Alternatively, $d_{\mathrm{B}}(K) \le \gamma$ whenever $\theta^\gamma M < 1$. This formulation will be useful in Exercise 12.5.

Now fix η with $0 < \eta < 1 - \alpha$, and let $r_0 = r_0(\eta)$. Cover K with $N(K, r_0)$ balls of radius r_0, $\{B(x_j, r_0)\}_{j=1}^N$, with centres $x_j \in K$. Apply f to every element of this cover. Since $f(K) = K$, this provides a new cover of K, $\{f(B(x_j, r_0))\}_{j=1}^N$. It follows from (12.3) that each of these images can be covered by M balls of radius $(\alpha + \eta)r_0$, ensuring that

$$N(K, (\alpha + \eta)r_0) \leq MN(K, r_0).$$

Applying this argument k times implies that

$$N(K, (\alpha + \eta)^k r_0) \leq M^k(K, r_0),$$

which via Lemma 3.2 (on taking the lim sup through a geometric sequence) yields

$$d_B(K) \leq \frac{\log M}{-\log(\alpha + \eta)}.$$

Since $\eta > 0$ was arbitrary we obtain (12.2). □

The key to applying this approach is to be able to prove (12.1), i.e. to find a way of estimating the number of balls of radius α required to cover $\mathrm{D}f(x)B(0,1)$. When $\mathrm{D}f(x)$ is the sum of a compact map and a contraction, we reduce the problem of covering $\mathrm{D}f(x)[B(0,1)]$ to the problem of covering $\mathrm{D}f(x)[B_Z(0,1)]$, where Z is some finite-dimensional subspace of $\mathscr{B}$. We then prove a covering result for balls in finite-dimensional subspaces. If $\mathrm{D}f(x)$ is the sum of a compact map and a contraction for every $x \in K$ in some suitably uniform way we can then obtain (12.1) with the same α and M for every $x \in K$.

12.1 Coverings of $T[B(0,1)]$ via finite-dimensional approximations

We want to cover the image of a ball under a linear map using balls of smaller radius. In order to do this we show that, given a linear map T that is the sum of a compact map and a contraction, $T[B(0,1)]$ can be well approximated by $T[B_Z(0,1)]$, where Z is some finite-dimensional subspace of $\mathscr{B}$.

We denote by $\mathscr{L}(\mathscr{B})$ the space of bounded linear transformations from $\mathscr{B}$ into itself, by $\mathcal{K}(\mathscr{B})$ the closed subspace of $\mathscr{L}(\mathscr{B})$ consisting of all compact linear transformations from $\mathscr{B}$ into itself, and define

$$\mathscr{L}_\lambda(\mathscr{B}) = \{T \in \mathscr{L}(\mathscr{B}) : T = L + C, \text{ with } C \in \mathcal{K}(\mathscr{B}) \text{ and } \|L\|_{\mathscr{L}(\mathscr{B})} < \lambda\}.$$

The result of the following lemma allows us to define a quantity $\nu_\lambda(T)$, which measures the 'effective dimension' of the range of T if we allow approximation to within a distance λ. Note that if T has finite rank n then $\nu_\lambda(T) = n$ for all $\lambda > 0$.

Lemma 12.2 *Let $\mathscr{B}$ be a Banach space and $T \in \mathscr{L}_{\lambda/2}(\mathscr{B})$. Then there exists a finite-dimensional subspace Z of $\mathscr{B}$ such that*

$$\mathrm{dist}(T[B(0,1)], T[B_Z(0,1)]) < \lambda. \tag{12.4}$$

We denote by $\nu_\lambda(T)$ the minimum $n \in \mathbb{N}$ such that (12.4) holds for some n-dimensional subspace of $\mathscr{B}$.

Proof Write $T = L + C$, where $C \in \mathcal{K}(\mathscr{B})$ and $L \in \mathscr{L}(\mathscr{B})$ is chosen such that $\|L\|_{\mathscr{L}(\mathscr{B})} < \lambda/2$. We show first that for any $\epsilon > 0$ there is a finite-dimensional subspace Z such that

$$\mathrm{dist}(C[B(0,1)], C[B_Z(0,1)]) < \epsilon.$$

Suppose that this is not the case. Choose some $x_1 \in \mathscr{B}$ with $\|x_1\| = 1$, and let $Z_1 = \mathrm{span}\{x_1\}$. Then

$$\mathrm{dist}(C[B(0,1)], C[B_{Z_1}(0,1)]) \geq \epsilon,$$

and so there exists an $x_2 \in \mathscr{B}$ with $\|x_2\| = 1$ such that

$$\|Cx_2 - Cx_1\| \geq \epsilon.$$

With $Z_2 = \mathrm{span}\{x_1, x_2\}$, one can find an x_3 with $\|x_3\| = 1$ such that

$$\|Cx_3 - Cx_1\| \geq \epsilon \quad \text{and} \quad \|Cx_3 - Cx_2\| \geq \epsilon.$$

Continuing inductively one can construct in this way a sequence $\{x_j\}$ with $\|x_j\| = 1$ such that

$$\|Cx_i - Cx_j\| \geq \epsilon \qquad i \neq j,$$

contradicting the compactness of C.

Now let $\tilde{\lambda} < \lambda$ be such that $2\|L\|_{\mathscr{L}(\mathscr{B})} < \tilde{\lambda} < \lambda$, and choose Z using the above argument so that

$$\mathrm{dist}(C[B(0,1)], C[B_Z(0,1)]) < \lambda - \tilde{\lambda}.$$

If $x \in B(0,1)$ and $z \in B_Z(0,1)$, then

$$\|Tx - Tz\| \leq \|L(x-z)\| + \|Cx - Cz\| \leq \tilde{\lambda} + \|Cx - Cz\|.$$

Hence,

$$\begin{aligned}\operatorname{dist}(T[B(0,1)], T[B_Z(0,1)]) &\le \tilde{\lambda} + \operatorname{dist}(C[B(0,1)], C[B_Z(0,1)]) \\ &< \lambda,\end{aligned}$$

and this completes the proof. □

We now need to be able to cover $T[B_Z(0, 1)]$ with $\mathscr{B}$-balls of a smaller radius. Since

$$T[B_Z(0,1)] \subseteq B_{T(Z)}(0, \|T\|),$$

we consider coverings of a ball in a general finite-dimensional subspace U of $\mathscr{B}$ with $\mathscr{B}$-balls of a smaller radius.

It is easy to estimate the number of balls required to cover a ball in $\mathbb{R}^n_\infty$ ($\mathbb{R}^n$ equipped with the ℓ^∞ norm) with balls of smaller radius, so we find a linear isomorphism between $\mathbb{R}^n_\infty$ and U which allows us to translate a covering in $\mathbb{R}^n_\infty$ to a covering in U.

Lemma 12.3 *If U is an n-dimensional subspace of a real Banach space X, then*

$$N(B_U(0,r), \rho) \le (n+1)^n \left(\frac{r}{\rho}\right)^n \qquad 0 < \rho \le r,$$

where the balls in the cover can be taken to have centres in U.

Proof First we find a linear isomorphism $J : \mathbb{R}^n_\infty \to U$ such that

$$\|J\|_{\mathscr{L}(\mathbb{R}^n_\infty, U)} \le n \qquad \text{and} \qquad \|J^{-1}\|_{\mathscr{L}(U, \mathbb{R}^n_\infty)} \le 1. \tag{12.5}$$

Let $\{x_1, \ldots, x_n\}$ be an Auerbach basis for U, and $\{f_1, \ldots, f_n\}$ the corresponding basis for U^*, i.e. $\|x_i\|_U = \|f_i\|_{U^*} = 1$ for $i = 1, \ldots, n$ and $f_i(x_j) = \delta_{ij}$, $i, j = 1, \ldots, n$ (for the existence of such a basis see Exercise 7.3). Define a map $J : \mathbb{R}^n_\infty \to U$ by setting

$$J(z) = \sum_{j=1}^n z_j x_j,$$

where $z = (z_1, \ldots, z_n)$. Then

$$\|J(z)\|_U = \left\| \sum_{j=1}^n z_j x_j \right\|_U \le \sum_{j=1}^n |z_j| \le n\|z\|_\infty,$$

which gives the first inequality in (12.5). On the other hand, if $x \in U$ with $x = \sum_{j=1}^{n} z_j x_j$ and $\|x\|_U \leq 1$ then since $z_j = f_j(x)$,

$$\|J^{-1}(x)\|_\infty = \|z\|_\infty = \max_{j=1,\ldots,n} |z_j| = \max_{j=1,\ldots,n} |f_j(x)| \leq \|x\|_U,$$

which yields the second inequality in (12.5).

Now since

$$B_U(0, r) = JJ^{-1}(B_U(0, r)) \subseteq J(B_{\mathbb{R}^n_\infty}(0, \|J^{-1}\|r)),$$

and $B_{\mathbb{R}^n_\infty}(0, \|J^{-1}\| r)$ can be covered by

$$\left(1 + \frac{\|J^{-1}\|r}{\rho/\|J\|}\right)^n = \left(1 + \|J\|\|J^{-1}\| \frac{r}{\rho}\right)^n \leq \left(1 + n\frac{r}{\rho}\right)^n \leq (n+1)^n \left(\frac{r}{\rho}\right)^n$$

balls in $\mathbb{R}^n_\infty$ of radius $\rho/\|J\|$ with centres in U, it follows that $B_U(0, r)$ can be covered by the same number of U-balls of radius ρ. □

Combining these two results we obtain the following corollary.

Corollary 12.4 *If $\mathscr{B}$ is a Banach space and $T \in \mathscr{L}_{\lambda/2}(\mathscr{B})$ then*

$$N(T[B(0,1)], 2\lambda) \leq \left[(n+1)\frac{\|T\|}{\lambda}\right]^n,$$

where $n = \nu_\lambda(T)$.

Proof Using Lemma 12.2 there is an n-dimensional subspace Z of $\mathscr{B}$ such that

$$\text{dist}(T[B(0,1)], T[B_Z(0,1)]) < \lambda. \tag{12.6}$$

Noting that $T(Z)$ is also an at most n-dimensional subspace of $\mathscr{B}$, one can use Lemma 12.3 to cover the ball $B_{T(Z)}(0, \|T\|)$ with balls $B(y_i, \lambda)$, $1 \leq i \leq k$, such that $y_i \in B(0, \|T\|)$ for each i and

$$k \leq \left[(n+1)\frac{\|T\|}{\lambda}\right]^n.$$

Thus

$$T[B_Z(0,1)] \subseteq B_{T(Z)}(0, \|T\|) \subseteq \bigcup_{i=1}^{k} B(y_i, \lambda). \tag{12.7}$$

We complete the proof by showing that

$$\bigcup_{i=1}^{k} B(y_i, 2\lambda) \supseteq T[B(0,1)].$$

If $x \in B(0, 1)$, it follows from (12.6) that there is a $y \in T[B_Z(0, 1)]$ such that $\|Tx - y\| < \lambda$. Since $y \in T[B_Z(0, 1)]$, it follows from (12.7) that $\|y - y_i\| \leq \lambda$ for some $i \in \{1, \ldots, k\}$, and so

$$\|Tx - y_i\| \leq \|Tx - y\| + \|y - y_i\| < 2\lambda,$$

i.e. $Tx \in B(y_i, 2\lambda)$. □

12.2 A dimension bound when $\mathrm{D}f \in \mathscr{L}_{\lambda/2}(\mathscr{B})$, $\lambda < \frac{1}{2}$

We now show, following Mañé (1981), that if $\mathrm{D}f(x) \in \mathscr{L}_{\lambda/2}$ for every $x \in K$ for some λ with $0 < \lambda < \frac{1}{2}$ then we have sufficient control to bound the dimension of K.

Theorem 12.5 *Let $\mathscr{B}$ be a Banach space, $U \subset \mathscr{B}$ an open set, and let $f : U \to \mathscr{B}$ be a continuously differentiable map. Suppose that $K \subset U$ is a compact set and assume that for some λ with $0 < \lambda < \frac{1}{2}$,*

$$\mathrm{D}f(x) \in \mathscr{L}_{\lambda/2}(\mathscr{B}) \qquad \textit{for all} \quad x \in K.$$

Then $n = \sup_{x \in K} \nu_\lambda(\mathrm{D}f(x))$ and $D = \sup_{x \in K} \|\mathrm{D}f(x)\|$ are finite, and

$$d_{\mathrm{B}}(K) \leq n \left\{ \frac{\log((n+1)D/\lambda)}{-\log(2\lambda)} \right\}.$$

Proof First we show that $n = \sup_{x \in K} \nu_\lambda(\mathrm{D}f(x))$ is finite. For each $x \in K$, there exists a finite-dimensional linear subspace Z_x such that

$$\mathrm{dist}(\mathrm{D}f(x)[B(0, 1)], \mathrm{D}f(x)[B_{Z_x}(0, 1)]) < \lambda.$$

Since $\mathrm{D}f(\cdot)$ is continuous, it follows that there exists a $\delta_x > 0$ such that

$$\mathrm{dist}(\mathrm{D}f(y)[B(0, 1)], \mathrm{D}f(y)[B_{Z_x}(0, 1)]) < \lambda$$

for all $y \in B(x, \delta_x)$, i.e. $\nu_\lambda(y) \leq \nu_\lambda(x)$ for all such y. The open cover of K formed by the union of $B(x, \delta_x)$ over x has a finite subcover, whence it follows that $n < \infty$.

Now, since $n = \sup_{x \in K} \nu_\lambda(\mathrm{D}f(x)) < \infty$, we can use Corollary 12.4 to deduce that

$$N(\mathrm{D}f(x)[B(0, 1)], 2\lambda) \leq \left[(n+1) \frac{D}{\lambda} \right]^n \qquad \text{for all} \quad x \in K.$$

The bound on the dimension now follows using Lemma 12.1. □

12.3 Finite dimension when $Df \in \mathscr{L}_1(X)$

The following corollary can be found in Hale, Magalhães, & Oliva (2002):

Corollary 12.6 *Suppose that $\mathscr{B}$ is a Banach space, $U \subset \mathscr{B}$ an open set, and $f : U \to \mathscr{B}$ a continuously differentiable map. Suppose that $K \subset U$ is a compact set such that $f(K) = K$, and that $Df(x) \in \mathscr{L}_1(\mathscr{B})$ for all $x \in K$. Then $d_B(K) < \infty$.*

Proof It follows from an argument similar to that used in the proof of Theorem 12.5 to show that $n < \infty$ that in fact there exists $\alpha < 1$ such that $Df(x) \in \mathscr{L}_\alpha(\mathscr{B})$ for all $x \in K$. Note that

$$D[f^p] = Df(f^{p-1}(x)) \circ \cdots \circ Df(x),$$

and that if $C_i \in \mathcal{K}(\mathscr{B})$ and $L_i \in \mathscr{L}(\mathscr{B})$, $i = 1, 2$, then

$$(C_1 + L_1) \circ (C_2 + L_2) = \underbrace{[C_1 \circ C_2 + C_1 \circ L_2 + L_1 \circ C_2]}_{\in \mathcal{K}(\mathscr{B})} + L_1 \circ L_2.$$

It follows that if $Df(x) \in \mathscr{L}_\alpha(\mathscr{B})$ with $\alpha < 1$ then $[D(f^p)](x) \in \mathscr{L}_{\alpha^p}(\mathscr{B})$. Thus for p large enough, $D(f^p)(x) \in \mathscr{L}_\lambda$ for some $\lambda < 1/4$, for every $x \in K$. One can now apply Theorem 12.5 to f^p in place of f (noting that $f^p(K) = K$) to deduce that $d_B(K) < \infty$. □

12.4 Semilinear parabolic equations in Hilbert spaces

We now prove a general result for the abstract semilinear parabolic equations we considered in Section 10.3.

Corollary 12.7 *Consider the semilinear parabolic equation*

$$du/dt = -Au + g(u) \qquad \text{with} \qquad u(0) = u_0 \in D(A^\alpha), \tag{12.8}$$

where $\alpha < 1$ and $g : D(A^\alpha) \to H$ is continuously differentiable. If this equation has a global attractor $\mathscr{A}$ that is bounded in $D(A^\alpha)$, then $d_B(\mathscr{A})$ is finite, where the dimension is measured in $D(A^\alpha)$.

With a little more work one can obtain an explicit bound on the dimension of $\mathscr{A}$, see Exercise 12.3.

Proof For $u_0 \in \mathscr{A}$, $S(t)u_0$ is given by the variation of constants formula (10.6),

$$S(t)u_0 = e^{-At}u_0 + \int_0^t e^{-A(t-s)} g(S(s)u_0)\, ds;$$

the derivative of $S(t)$ with respect to u at u_0, $\mathrm{D}S(t;u_0)$, is an element of $\mathscr{L}(D(A^\alpha))$ and satisfies

$$\mathrm{D}S(t;u_0)=\mathrm{e}^{-At}+\int_0^t \mathrm{e}^{-A(t-s)}\mathrm{D}g(S(s)u_0)\mathrm{D}S(s;u_0)\,\mathrm{d}s$$

(see Henry (1981, Theorem 3.4.4)). Using the bound on $\|A^\gamma \mathrm{e}^{-At}\|_{\mathscr{L}(H)}$ in (10.9) we obtain

$$\begin{aligned}\|\mathrm{D}S(t;u_0)\|_{\mathscr{L}(D(A^\alpha))} &\le \|\mathrm{e}^{-At}\|_{\mathscr{L}(D(A^\alpha))}\\ &\quad+\int_0^t \|A^\alpha \mathrm{e}^{-A(t-s)}\|_{\mathscr{L}(H)}\|\mathrm{D}g(S(s(u_0)))\|_{\mathscr{L}(D(A^\alpha),H)}\|\mathrm{D}S(s;u_0)\|_{\mathscr{L}(D(A^\alpha))}\\ &\le 1+cM\int_0^t (t-s)^{-\alpha}\|\mathrm{D}S(s;u_0)\|_{\mathscr{L}(D(A^\alpha))}\,\mathrm{d}s,\end{aligned}$$

where

$$M=\sup\{\|\mathrm{D}g(x)\|_{\mathscr{L}(D(A^\alpha),H)}:x\in\mathscr{A}\}. \tag{12.9}$$

It follows from this inequality, using the result of Exercise 12.2, that

$$\|\mathrm{D}S(t;u_0)\|_{\mathscr{L}(D(A^\alpha))}\le K:=2\mathrm{e}^{[2cM\Gamma(1-\alpha)]^{1/(1-\alpha)}} \qquad \text{for all} \qquad t\in[0,1]. \tag{12.10}$$

Now choose $\epsilon>0$ such that $\alpha+\epsilon<1$. Taking advantage of (12.10) one can use very similar estimates to those above to show that

$$\|\mathrm{D}S(t)\|_{\mathscr{L}(D(A^\alpha),D(A^{\alpha+\epsilon}))}\le ct^{-\epsilon}+cMK\int_0^t (t-s)^{-(\alpha+\epsilon)}\,\mathrm{d}s.$$

Since $D(A^{\alpha+\epsilon})$ is compactly embedded in $D(A^\alpha)$, this shows that $\mathrm{D}S(t)$ is compact for any $t>0$. That $d_{\mathrm{B}}(\mathscr{A})$ is finite now follows immediately from Corollary 12.6 applied with $\mathscr{B}=D(A^\alpha)$. □

This approach is also applicable to the two-dimensional Navier–Stokes equations. We saw in Section 10.4 that the Navier–Stokes equations can be cast in the form (12.8) with g locally Lipschitz from $D(A^\alpha)$ into H provided that $\alpha>1/2$, and Exercise 11.6 guarantees that if $f\in H$ then the attractor is bounded in $D(A)$, so $\mathscr{A}$ is certainly bounded in $D(A^{3/4})$ (to choose some fixed α with $1/2<\alpha<1$). Corollary 12.7 then implies that the dimension of $\mathscr{A}$ measured in $D(A^{3/4})$ is finite; so certainly the dimension of $\mathscr{A}$ measured in H is finite (see (3.5)).

As remarked at the beginning of this chapter, obtaining good bounds on the dimension of the Navier–Stokes attractor has been an active area of research. Of course, such bounds use the Hilbert space theory rather than the Banach space approach developed above. In the periodic case, Constantin, Foias, &

Temam (1988) showed that

$$d_{\rm B}(\mathscr{A}) \leq c\|f\|^{2/3}(1+\log\|f\|)^{1/3};$$

there are example forcing functions f for which the dimension is bounded below by $c'\|f\|^{2/3}$ (Liu, 1993), so this bound is essentially sharp. For a simplified proof and further discussion, see Doering & Gibbon (1995).

Exercises

12.1 Let $\mathscr{B}$ be a Banach space and assume that $f \in C^1(X)$, that K is a compact set such that $f(K)=K$, and that for every $x \in K$ the derivative $\mathrm{D}f(x)$ has finite rank $\nu(x)$ with $\sup_{x\in K}\nu(x) := \nu < \infty$. Show that $d_{\rm B}(K) \leq \nu$.

12.2 Suppose that $X(t)$ satisfies

$$X(t) \leq a + b\int_0^t (t-s)^{-\alpha}X(s)\,\mathrm{d}s.$$

With $K=(2b\Gamma(1-\alpha))^{1/(1-\alpha)}$ show that $Y(t)=2a\mathrm{e}^{Kt}$ satisfies

$$\dot{Y} \geq a + b\int_0^t (t-s)^{-\alpha}Y(s)\,\mathrm{d}s,$$

and hence that $X(t) \leq 2a\mathrm{e}^{Kt}$.

12.3 Recalling that $\Gamma(z)=\int_0^\infty t^{z-1}\mathrm{e}^{-t}\,\mathrm{d}t$, take up the argument of Corollary 12.7 immediately after (12.10), and show that

$$\|Q_n\mathrm{D}S(1;u_0)\|_{\mathscr{L}(D(A^\alpha))} \leq \mathrm{e}^{-\lambda_{n+1}} + \frac{cKM\Gamma(1-\alpha)}{(\lambda_{n+1}-1)^{1-\alpha}}, \qquad (12.11)$$

where P_n is the orthogonal projection onto the space spanned by the first n eigenfunctions of A and $Q_n = I - P_n$ is its orthogonal complement. (Choosing n large enough that the right-hand side is strictly less than $1/8$, one can then apply Theorem 12.5 with $\mathscr{B}=D(A^\alpha)$ and $\lambda = 1/4$ to deduce that

$$d_{\rm B}(\mathscr{A}) \leq n\,\frac{\log[4K(n+1)]}{\log 2}.$$

Dropping the first term in (12.11) we require $\lambda_{n+1}^{1-\alpha} \gtrsim cKM$, which using the estimate on K in (12.10) becomes $\lambda_{n+1} \gtrsim cM_\alpha\mathrm{e}^{cM_\alpha}$ with $M_\alpha = M^{1/(1-\alpha)}$. Assuming that $\lambda_n \sim n^\gamma$, this yields $n \lesssim M_\alpha^{1/\gamma}\mathrm{e}^{cM_\alpha}$, and hence $d_{\rm B}(\mathscr{A}) \lesssim M_\alpha^{1+(1/\gamma)}\mathrm{e}^{cM_\alpha}$.)

The remainder of the exercises in this chapter outline the more refined theory that is available to estimate the dimension of attractors in the Hilbert space case. Exercise 12.4 provides a good estimate on coverings of $T[B(0,1)]$ in terms of the singular values of T. Exercise 12.5 converts this into a bound on the dimension via Lemma 12.1, and the remaining exercises give an indication of how to apply this method efficiently in applications (which usually arise from continuous time systems).

The covering argument of Exercise 12.4 requires the following two results. The proof of the first is essentially the same as that of Lemma 14.2, below, and the proof of the second can be found in Chepyzhov & Vishik (2002, Lemma 2.2).

Lemma 12.8 *Let $T : H \to H$ be a compact linear map, and denote by T^* the Hilbert adjoint of T, i.e. the unique $T^* \in \mathscr{L}(H)$ such that $(Tx, y) = (x, T^*y)$ for every $x, y \in H$. Then $T[B(0,1)]$ is an ellipsoid whose semiaxes are $\{Te_j\}$, and $\|Te_j\| = \alpha_j$, where $\{e_j\}$ are the eigenvectors of T^*T corresponding to its non-zero eigenvalues α_j^2.*

Lemma 12.9 *Let E be an ellipsoid in H with semiaxes $\alpha_1 \ge \alpha_2 \ge \alpha_3 \ge \cdots$. Then for any $r < \alpha_1$, the number of balls of radius $\sqrt{2}r$ required to cover E is less than $4^j\omega_j/r^j$, where $\omega_j = \alpha_1 \cdots \alpha_j$ and j is the largest integer such that $r \le \alpha_j$.*

Now for a given compact $T : H \to H$, let $\alpha_j(T)$ denote the square roots of the eigenvalues of T^*T listed in decreasing order,

$$\alpha_1(T) \ge \alpha_2(T) \ge \alpha_3(T) \ge \cdots,$$

and set $\omega_n(T) = \alpha_1(T)\alpha_2(T)\cdots\alpha_n(T)$.

12.4 Let $U \subset H$ be an open set, $f : U \to H$ a continuously differentiable map, and K a compact subset of U that is invariant under f. Suppose that

$$\alpha_n(\mathrm{D}f(x)) \le \bar{\alpha}_n \qquad \text{and} \qquad \omega_n(\mathrm{D}f(x)) \le \bar{\omega}_n \qquad \text{for all} \qquad x \in K,$$

where $\bar{\alpha}_1 \ge \bar{\alpha}_2 \ge \cdots$ and $\bar{\alpha}_n^n \le \bar{\omega}_n$. Show that, for any choice of $d \in \mathbb{N}$,

$$N(\mathrm{D}f(u)[B(0,1)], \theta) \le M \qquad \text{for all} \qquad u \in K,$$

where

$$\theta = \sqrt{2}\bar{\omega}_d^{1/d} \qquad \text{and} \qquad M = \max_{1 \le j \le d} \frac{4^j \bar{\omega}_j}{\bar{\omega}_d^{j/d}}.$$

[Hint: for each $u \in K$, consider the two cases $\bar{\omega}_d^{1/d} < \alpha_1(\mathrm{D}f(u))$ and $\bar{\omega}_d^{1/d} \geq \alpha_1(\mathrm{D}f(u))$ separately.]

12.5 Under the same conditions as in the previous exercise, use Lemma 12.1 to show that if

$$\bar{\omega}_d < 1 \qquad \text{and} \qquad \bar{\omega}_d^{\gamma} \max_{1 \leq j \leq d} \frac{\bar{\omega}_j^d}{\bar{\omega}_d^j} < 1, \tag{12.12}$$

then $d_{\mathrm{B}}(K) \leq \gamma$. [Hint: consider f^k rather than f, for some k chosen sufficiently large. You may assume that $\omega_j(TS) \leq \omega_j(T)\omega_j(S)$; the proof of this uses the abstract theory of multilinear operators on Hilbert spaces, see Chapter V of Temam (1988), for example.]

12.6 Write $\bar{q}_j = \log \bar{\omega}_j$, and assume that $\bar{q}_j \leq q_j$, where q_j is a concave function of j. Show that $q_n < 0$ implies that $d_{\mathrm{B}}(K) \leq n$. (This observation is due to Chepyzhov & Ilyin (2004).)

Constantin & Foias (1985) showed that if $S(\cdot)$ is a semigroup on H arising from the differential equation $\mathrm{d}u/\mathrm{d}t = F(u)$, such that $\mathrm{D}S(t; u_0)$ is the solution of $\mathrm{d}U/\mathrm{d}t = F'(S(t)u_0)U$ with $U(0) = \mathrm{id}$, then

$$q_n(\mathrm{D}S(t; u_0)) \leq \int_0^t \mathrm{Tr}_n(F'(u(s)))\,\mathrm{d}s, \tag{12.13}$$

where

$$\mathrm{Tr}_n(L) = \sup \left\{ \sum_{j=1}^{n} (\psi_j, L\psi_j) : \ \{\psi_j\}_{j=1}^n \text{ are orthonormal in } H \right\}.$$

12.7 Let A be an unbounded self-adjoint positive operator with compact inverse, with eigenvalues $\{\lambda_j\}_{j=1}^{\infty}$ arranged in nondecreasing order. Show that for any choice of n orthonormal elements $\{\phi_j\}_{j=1}^n$ in H,

$$\sum_{j=1}^{n} \|A^{1/2}\phi_j\|^2 = \sum_{j=1}^{n} (\phi_j, A\phi_j) \geq \sum_{j=1}^{n} \lambda_j.$$

12.8 Consider a semigroup $S(\cdot)$ defined on H that arises from the semilinear evolution equation

$$\mathrm{d}u/\mathrm{d}t = -Au + g(u), \tag{12.14}$$

where A is as in the previous exercise, $0 \leq \alpha \leq \frac{1}{2}$, and $g : D(A^{\alpha}) \to H$ is continuously differentiable. Assuming that $\mathrm{D}S(t; u_0)$ is the solution of

$$\mathrm{d}U/\mathrm{d}t = -AU + \mathrm{D}g(u(t))U \qquad \text{with} \qquad U(0) = \mathrm{id},$$

show that $d_{\mathrm{B}}(\mathscr{A}) \leq n$ provided that

$$\frac{1}{n}\sum_{j=1}^{n} \lambda_j > M^{1/(1-\alpha)}, \tag{12.15}$$

where M is defined in (12.9). (If we assume that $\lambda_n \sim n^{\gamma}$, then we have $\sum_{j=1}^{n} \lambda_n \sim n^{1+\gamma}$ and $d_{\mathrm{B}}(\mathscr{A}) \lesssim M_\alpha^{1/\gamma}$; compare this with the bound from Exercise 12.3 (using the Banach space method), which was exponential in M_α. However, note that here we have made the additional assumption that (12.14) makes sense for $u_0 \in H$.)

13

Thickness exponents of attractors

Ott, Hunt, & Kaloshin (2006) conjectured that 'many of the attractors associated with the evolution equations of mathematical physics have thickness exponent zero'.

In this chapter we give two results in this direction. The first, due to Friz & Robinson (1999), shows that in some sense the thickness exponent is 'inversely proportional to smoothness': if $U \subset \mathbb{R}^m$ and $\mathscr{A}$ is a subset of $L^2(U)$ that is bounded in the Sobolev space $H^s(U)$ then $\tau(\mathscr{A}) \le m/s$, where the thickness of $\mathscr{A}$ is measured in $L^2(U)$. So if an attractor is 'smooth' (i.e. is bounded in $H^s(\Omega)$ for every s) then it has zero thickness exponent.

The second result, due to Pinto de Moura & Robinson (2010c), is closer in spirit to the above conjecture. This shows that the attractors of equations that can be written as semilinear parabolic equations

$$\mathrm{d}u/\mathrm{d}t = -Au + g(u) \tag{13.1}$$

have zero Lipschitz deviation. The argument is related to a backwards uniqueness property for solutions of (13.1), whose proof is due to Kukavica (2007).

13.1 Zero thickness

We begin with an 'analytical' result which does not rely on the dynamics associated with the set X or the form of the underlying equations, but only makes assumptions on the smoothness of functions that make up X.

Lemma 13.1 *Let $U \subset \mathbb{R}^m$ be a smooth bounded domain. Let X be a compact subset of $[L^2(U)]^n$ such that*

$$\sup_{u \in X} \|u\|_{[H^s(U)]^n} < \infty$$

for some $s \geq 1$. Then $\tau(X) \leq m/s$.

We will use a similar argument to what follows for Lemma 15.5. The proof is due to Friz & Robinson (1999), see also Robinson (2008).

Proof A proof for the case $n = 1$ is sufficient; if $n > 1$ the argument can be applied to each component of the functions in X. Let U', U'' be smooth bounded domains such that

$$\overline{U} \subset U' \qquad \text{and} \qquad \overline{U'} \subset U''$$

with both inclusions strict. Let $E : H^s(U) \to H^s(U'')$ be a bounded extension operator such that for all $u \in H^s(U)$,

$$\|E[u]\|_{H^s(U'')} \leq C\|u\|_{H^s(U)} \tag{13.2}$$

and the support of $E[u]$ is contained in U' (e.g. Theorem 7.25 in Gilbarg & Trudinger (1983)).

Let A denote the Laplacian operator on U'', with Dirichlet boundary conditions ($u = 0$ on $\partial U''$). The Laplacian on such a domain has a sequence $\{w_j\}$ of eigenfunctions with corresponding eigenvalues λ_j ($Aw_j = \lambda_j w_j$) which, if ordered so that $\lambda_{j+1} \geq \lambda_j$, satisfy $\lambda_j \sim j^{2/m}$ (see Davies (1995), for example).

Since, for any $u \in X$, the support of $E[u]$ is contained in U', $E[u] \in D(A^{s/2})$. It follows from (13.2) and the inequality

$$\|A^{s/2}v\|_{L^2(U'')} \leq \|v\|_{H^s(U'')} \qquad \text{for all} \qquad v \in D(A^{s/2})$$

that $E[X]$ $(= \bigcup_{u \in X} E[u])$ is uniformly bounded in $D(A^{s/2})$.

Now define the orthogonal projection P_k onto the space spanned by the first k eigenfunctions of A,

$$P_k v = \sum_{j=1}^{k} (v, w_j) w_j,$$

and its orthogonal complement $Q_k = I - P_k$. Note that $\|Q_k A^{-s} u\| \leq \lambda_{k+1}^{-s} \|u\|$.

Consider the approximation of u in the k-dimensional subspace spanned by $\{w_j|_U\}_{j=1}^k$ given by $(P_k E[u])|_U$:

$$\begin{aligned}\|u - (P_k E[u])|_U\|_{L^2(U)} &\le \|E[u] - P_k E[u]\|_{L^2(U'')}\\ &= \|Q_k E[u]\|_{L^2(U'')}\\ &= \|Q_k A^{-s/2} A^{s/2} E[u]\|_{L^2(U'')}\\ &\le \|Q_k A^{-s/2}\|_{\mathscr{L}(H)} \|A^{s/2} E[u]\|_{L^2(U'')}\\ &\le \lambda_{k+1}^{-s/2} \|E[u]\|_{H^s(U'')}\\ &\le C\lambda_{k+1}^{-s/2} \|u\|_{H^s(U)}\\ &\le K\|u\|_{H^s(U)} k^{-s/m},\end{aligned}$$

for some constant K, and so $\tau(X) \le m/s$ using Exercise 7.1. □

Clearly if an attractor is 'smooth', i.e. bounded in $H^s(\Omega)$ for every s, then its thickness is zero. For the two-dimensional Navier–Stokes equations this can be translated to an assumption on the smoothness of the forcing term f.

Corollary 13.2 *The attractor of the two-dimensional Navier–Stokes equations has zero thickness if* $f \in C^\infty(\Omega)$.

Proof Guillopé (1982) showed that if $f \in C^\infty(\Omega)$ then the attractor is bounded in $H^k(\Omega)$ for every $k \in \mathbb{N}$, and the result follows immediately from Lemma 13.1. □

13.2 Zero Lipschitz deviation

While there are currently no examples of attractors of natural models that have been proved to have nonzero thickness, there is no proof available that 'many attractors' do have zero thickness exponent.

We now show that the Lipschitz deviation of a large class of attractors is zero. We work with semilinear parabolic equations of the form

$$\mathrm{d}u/\mathrm{d}t = -Au + g(u) \tag{13.3}$$

with g Lipschitz from $D(A^\alpha)$ into H when restricted to $\mathscr{A}$:

$$\|g(u) - g(v)\| \le L\|u - v\|_\alpha \quad \text{whenever} \quad u, v \in \mathscr{A}, \tag{13.4}$$

cf. Section 10.3. We make the additional assumption that $0 \le \alpha \le \frac{1}{2}$.

Abstract existence and uniqueness results for (13.3) under the condition that g is Lipschitz from $D(A^\alpha)$ into H require the initial condition to be in $D(A^\alpha)$, and generate a dynamical system on $D(A^\alpha)$. However, here we consider the attractor $\mathscr{A}$ as a subset of H, and show that the Lipschitz deviation of $\mathscr{A}$ as measured in H is zero. As remarked in Section 10.4, which treated the particular example of the two-dimensional Navier–Stokes equations, one can often obtain existence and uniqueness results in larger spaces (like H) using equation-specific methods, and then employ the abstract formulation to deduce further properties of these solutions (since for $t > 0$ they will be smooth enough for the abstract theory to apply, i.e. $u(t) \in D(A^\alpha)$); this is the approach we adopt here.

Theorem 13.3 *Take $\alpha \in [0, \frac{1}{2}]$, and suppose that (13.3) has an attractor $\mathscr{A}$ that is bounded in $D(A^\alpha)$. Pick M_0 such that $\mathscr{A} \subset B_H(0, M_0/8)$.*

(i) *There exists a constant $C > 0$ such that*

$$\frac{\|A^{1/2}(u-v)\|^2}{\|u-v\|^2 \log(M_0^2/\|u-v\|^2)} < C \qquad \text{for all} \qquad u, v \in \mathscr{A}.$$

(ii) *For each $n \in \mathbb{N}$ there exists a 1-Lipschitz function $\Phi_n : P_n H \to Q_n H$, $Q_n = I - P_n$,*

$$\|\Phi_n(p) - \Phi_n(\bar{p})\| \le \|p - \bar{p}\| \qquad \text{for all} \qquad p, \bar{p} \in P_n H,$$

such that

$$\operatorname{dist}(\mathscr{A}, G_{P_n H}[\Phi_n]) \le M_0^2 e^{-\lambda_{n+1}/2C}. \tag{13.5}$$

(iii) *If in addition*

$$\lim_{n\to\infty} \frac{\lambda_n}{\log n} = \infty \tag{13.6}$$

then $\operatorname{dev}(\mathscr{A}) = 0$ *(where the Lipschitz deviation is measured in H).*

The proof of (i) is due to Kukavica (2007); Pinto de Moura & Robinson (2010c) show that (ii) is a consequence of (i) (using an argument of Foias, Manley, & Temam (1988)) and observe that (iii) follows from (ii) (see also Pinto de Moura & Robinson (2010b).)

The condition (13.6) will be satisfied in most interesting examples, since for an elliptic operator of order $2p$ defined in $\Omega \subset \mathbb{R}^m$, $\lambda_n \sim n^{2p/m}$ (see Davies (1995), for example).

Proof (i) Let $w(t) = u(t) - v(t)$, $L(t) = \log(M_0^2/\|w(t)\|^2)$ (so in particular $L \ge 1$), and set

$$Q(t) = \frac{\|A^{1/2}w\|^2}{\|w\|^2} \qquad \text{and} \qquad \tilde{Q}(t) = \frac{Q(t)}{L(t)}.$$

With $h(t) = g(u(t)) - g(v(t))$, w satisfies

$$\frac{\mathrm{d}w}{\mathrm{d}t} = -Aw + h(t). \tag{13.7}$$

Then (cf. Constantin, Foias, Nicolaenko, and Temam (1989), or Lemma III 6.1 in Temam (1988))

$$\begin{aligned}\frac{1}{2}\frac{\mathrm{d}Q}{\mathrm{d}t} &= \frac{(A^{1/2}w_t, A^{1/2}w)}{\|w\|^2} - \frac{(w, w_t)\|A^{1/2}w\|^2}{\|w\|^4} \\ &= \frac{(-Aw + h, Aw)}{\|w\|^2} - (w, -Aw + h)\frac{Q(t)}{\|w\|^2} \\ &= -\|A\hat{w}\|^2 + Q(t)^2 + \left(\frac{h}{\|w\|}, (A - Q(t)I)\hat{w}\right),\end{aligned}$$

where $\hat{w} = w/\|w\|$. Noting that

$$\begin{aligned}\|(A - Q(t)I)\hat{w}\|^2 &= \|A\hat{w}\|^2 - 2Q(t)(A\hat{w}, \hat{w}) + Q(t)^2\|\hat{w}\|^2 \\ &= \|A\hat{w}\|^2 - Q(t)^2,\end{aligned}$$

it follows that

$$\frac{1}{2}\frac{\mathrm{d}}{\mathrm{d}t}Q + \|(A - Q(t)I)\hat{w}\|^2 = \left(\frac{h}{\|w\|}, (A - Q(t)I)\hat{w}\right).$$

Since

$$\frac{\mathrm{d}\tilde{Q}}{\mathrm{d}t} = L^{-1}\frac{\mathrm{d}Q}{\mathrm{d}t} - QL^{-2}\frac{\mathrm{d}L}{\mathrm{d}t}$$

and

$$\frac{\mathrm{d}L}{\mathrm{d}t} = 2Q(t) - 2\frac{(h, w)}{\|w\|^2},$$

we can obtain

$$\frac{1}{2}\frac{\mathrm{d}\tilde{Q}}{\mathrm{d}t} + \tilde{Q}^2 + \frac{\|(A - QI)\hat{w}\|^2}{L} = \frac{(h, (A - QI)\hat{u})}{\|w\|L} + \frac{(h, w)\tilde{Q}}{\|w\|^2 L}.$$

The right-hand side is bounded by

$$\begin{aligned}&\frac{\|h\|\|(A - QI)\hat{w}\|}{\|w\|L} + \frac{\|h\|\tilde{Q}}{\|w\|L} \\ &\quad \le \frac{1}{2}\frac{\|h\|^2}{\|w\|^2 L} + \frac{1}{2}\frac{\|(A - QI)\hat{w}\|^2}{L^k} + \frac{1}{2}\tilde{Q}^2 + \frac{1}{2}\frac{k\|h\|^2}{\|w\|^2 L^2};\end{aligned}$$

since $L \ge 1$, we obtain

$$\frac{\mathrm{d}\tilde{Q}}{\mathrm{d}t} + \tilde{Q}^2 + \frac{\|(A - QI)\hat{w}\|^2}{L} \le \frac{2\|h\|^2}{\|w\|^2 L}. \tag{13.8}$$

Now we use the fact that g is locally Lipschitz (10.5) and that $\mathscr{A}$ is bounded in $D(A^\alpha)$ to deduce that

$$\|h(t)\| = \|g(u(t)) - g(v(t))\| \leq K\|u(t) - v(t)\|_\alpha = K\|w(t)\|_\alpha$$

for some $K > 0$; combine this with the interpolation inequality $\|w\|_\alpha \leq \|w\|^{1-2\alpha}\|w\|_{1/2}^{2\alpha}$ (see Exercise 13.4) to bound the right-hand side:

$$\frac{2\|h\|^2}{\|w\|^2 L} \leq \frac{2K\|w\|_\alpha^2}{\|w\|_2 L} \leq \frac{2K\|w\|_{1/2}^{4\alpha}}{\|w\|^{4\alpha} L} = \frac{2K}{L^{1-2\alpha}}\tilde{Q}^{2\alpha} \leq 2K\tilde{Q}^{2\alpha} \tag{13.9}$$

since $L \geq 1$ and $\alpha \in [0, \frac{1}{2}]$. Dropping the third term on the left-hand side of (13.8) and using Young's inequality ($2K\tilde{Q}^{2\alpha} \leq \frac{1}{2}\tilde{Q}^2 + K'$),

$$\frac{d\tilde{Q}}{dt} + \frac{1}{2}\tilde{Q}^2 \leq K'. \tag{13.10}$$

Finally the result of Exercise 13.2 guarantees that whatever the value of $\tilde{Q}(0)$, (13.10) implies that

$$\tilde{Q}(1) \leq 2 + (2K')^{1/2}.$$

Since $\mathscr{A}$ is invariant, given $u, v \in \mathscr{A}$ there exist $u_0, v_0 \in \mathscr{A}$ such that $u = S(1)u_0$ and $v = S(1)v_0$. It follows that

$$\frac{\|A^{1/2}(u - v)\|^2}{\|u - v\|^2 \log(M_0^2/\|u - v\|^2)} \leq C \tag{13.11}$$

for all $u, v \in \mathscr{A}$.

(ii) Take $u, v \in \mathscr{A}$ and set $w = u - v$. Writing $w = P_n w + Q_n w$, observe that

$$\begin{aligned}\|A^{1/2}w\|^2 &= \|A^{1/2}(P_n w + Q_n w)\|_{L^2}^2 \\ &= \|A^{1/2}P_n w\|^2 + \|A^{1/2}Q_n w\|^2 \\ &\geq \lambda_{n+1}\|Q_n w\|^2.\end{aligned}$$

It follows from (13.11) that

$$\|A^{1/2}w\|^2 \leq C\big(\|P_n w\|^2 + \|Q_n w\|^2\big)\log\big(M_0^2/\|Q_n w\|^2\big). \tag{13.12}$$

Now consider a subset X of $\mathscr{A}$ that is maximal for the relation

$$\|Q_n(u - v)\| \leq \|P_n(u - v)\| \qquad \text{for all} \qquad u, v \in X. \tag{13.13}$$

For every $p \in P_n X$ with $p = P_n u$, $u \in X$, define $\phi_n(p) = Q_n u$; (13.13) shows that this is well defined, and that

$$\|\phi_n(p) - \phi_n(\bar{p})\| \leq \|p - \bar{p}\| \qquad \text{for all} \qquad p, \bar{p} \in P_n X.$$

Since X is a closed subset of the compact set $\mathscr{A}$, $P_n X$ is closed, and so $\phi_n : P_n X \to Q_n H$ can be extended to a function $\Phi_n : P_n H \to Q_n H$ (i.e. one defined on the whole of $P_n X$) with the same Lipschitz constant (see Wells & Williams (1975)).

If $u \in \mathscr{A}$ but $u \notin X$ then there is a $v \in X \subset \mathscr{A}$ such that

$$\|Q_n(u - v)\| > \|P_n(u - v)\|. \tag{13.14}$$

Since $\lambda_{n+1}\|Q_n w\|^2 \leq \|A^{1/2}w\|^2$ it follows from (13.12) that

$$\lambda_{n+1}\|Q_n w\|^2 \leq 2C\|Q_n w\|^2 \log(M_0^2/\|Q_n w\|^2),$$

and hence

$$\|Q_n w\|^2 \leq M_0^2 \mathrm{e}^{-\lambda_{n+1}/2C}.$$

The inequality (13.5) now follows using (13.14).

(iii) We want to apply the definition of the Lipschitz deviation, using

$$\delta_1(\mathscr{A}, \varepsilon_n) \leq n \qquad \text{with} \qquad \varepsilon_n = M_0^2 \mathrm{e}^{-\lambda_{n+1}/2C}$$

from (13.5). Following an argument similar to that used in Lemma 3.2, take $\epsilon > 0$ with $\varepsilon_{n+1} \leq \epsilon < \varepsilon_n$, and then

$$\frac{\log \delta_1(\mathscr{A}, \epsilon)}{-\log \epsilon} \leq \frac{\log \delta_1(\mathscr{A}, \varepsilon_{n+1})}{-\log \varepsilon_n} \leq \frac{\log(n+1)}{(\lambda_{n+1}/2C) - 2\log M_0}.$$

Then $\mathrm{dev}(\mathscr{A}) \leq \mathrm{dev}_1(\mathscr{A}) = 0$ provided that (13.6) holds. □

The increased power of this result over that of Lemma 13.1 is clearly demonstrated by the following consequence for the two-dimensional Navier–Stokes equations (recall that H is essentially the space L^2 of square integrable functions).

Corollary 13.4 *The attractor of the two-dimensional Navier–Stokes equations has zero Lipschitz deviation if $f \in H$.*

Proof We saw in Section 10.4 that the Navier–Stokes equations can be written in the form (13.3) where g satisfies

$$\|g(u) - g(v)\| \leq c_\alpha[\|u\|_\alpha + \|v\|_\alpha]\|u - v\|_{1/2}$$

for any $\alpha > 1/2$ (this was (10.20)). For $f \in H$ the attractor is bounded in $D(A)$ (see Exercise 11.6) and so it follows that $\|u\|_\beta$ is bounded for all $u \in \mathscr{A}$, for any $0 \leq \beta \leq 1$. Hence

$$\|g(u) - g(v)\| \leq C\|u - v\|_{1/2} \qquad \text{for all} \qquad u, v \in \mathscr{A},$$

i.e. g satisfies (13.4) with $\alpha = 1/2$. One can therefore apply Theorem 13.3, since the eigenvalues of the Stokes operator on a two-dimensional periodic domain satisfy $\lambda_n \sim cn$ (see Exercise 13.3). □

A first version of the result of part (iii) of Theorem 13.3 was proved in Pinto de Moura & Robinson (2010a), using the dynamical 'squeezing property' due to Eden *et al.* (1994), see Exercise 13.5. One could also appeal more directly to results on families of approximate inertial manifolds of exponential order due to Debussche & Temam (1994) and Rosa (1995).

Exercises

13.1 Suppose that A^β is Lipschitz on $\mathscr{A}$ for some $\beta > 0$:

$$\|A^\beta(u - v)\| \le L\|u - v\| \qquad \text{for all} \qquad u, v \in \mathscr{A}.$$

Show that for n sufficiently large

$$\|Q_n(u - v)\| \le \|P_n(u - v)\| \qquad \text{for all} \qquad u, v \in \mathscr{A},$$

and hence that $\mathscr{A}$ is contained in the graph of a 1-Lipschitz function over $P_n H$. (This immediately provides a bi-Lipschitz embedding of $\mathscr{A}$ into $\mathbb{R}^n$.)

13.2 Show that if $y \ge 0$ and $\dot{y} + \gamma y^2 \le \delta$ with $\gamma > 0$, $\delta \ge 0$, then

$$y(t) \le \left(\frac{\delta}{\gamma}\right)^{1/2} + \frac{1}{\gamma t} \tag{13.15}$$

for all $t \ge 0$. [Hint: consider $y(t) = z(t) + (\delta/\gamma)^{1/2}$ for $t \in [0, t_0]$ with t_0 chosen so that $z(t) \ge 0$.] A more general version of this result, due to Ghidaglia, can be found as Lemma III.5.1 in Temam (1988).

13.3 The eigenvalues of the Stokes operator on a two-dimensional periodic domain $[0, 2\pi]^2$ are the sums of two square integers. If $\{\lambda_n\}$ are these eigenvalues arranged in nondecreasing order, show that $\frac{1}{2}n \le \lambda_n \le 2n$.

13.4 Use Hölder's inequality and the definition of the fractional powers of A in (10.2) to show that if $u \in D(A^\beta)$ then for any $\alpha < \beta$

$$\|A^\alpha u\| \le \|u\|^{1-(\alpha/\beta)}\|A^\beta u\|^{\alpha/\beta}.$$

13.5 Let $S(t)$ be the semigroup generated by (10.4), and assume the existence of a global attractor $\mathscr{A} \subset B_H(0, M)$. Eden *et al.* (1994) show that there exists a time t^* and an n_0 such that for all $n \ge n_0$ there exists an orthogonal projection P_n of rank n such that for every $u, v \in \mathscr{A}$ either

$$\|Q_n(S(t^*)u - S(t^*)v)\| \leq \|P_n(S(t^*)u - S(t^*)v)\| \tag{13.16}$$

(where $Q_n = I - P_n$) or

$$\|S(t^*)u - S(t^*)v\| \leq \delta_n \|u - v\|, \tag{13.17}$$

with

$$\delta_n \leq c_0 \mathrm{e}^{-\sigma \lambda_{n+1}},$$

where c_0 and σ are constants depending only on L and α in (10.5). Use an argument similar to that employed above to prove part (ii) of Theorem 13.3 to show that for each $n \geq n_0$, $\mathscr{A}$ lies within a $4M\delta_n$ neighbourhood of a 1-Lipschitz graph over an n-dimensional subspace of H. (As in Theorem 13.3 it then follows that $\mathrm{dev}(\mathscr{A}) = 0$.) This result, the source of the argument for part (ii) of Theorem 13.3, can be found in Foias *et al.* (1988) and Robinson (2001, Proposition 14.2).

13.6 Find an $f \in L^2$ such that the attractor of the two-dimensional Navier–Stokes equations

$$\mathrm{d}u/\mathrm{d}t + Au + B(u, u) = f$$

is not bounded in H^3.

13.7 Show that part (i) of Theorem 13.3 implies that if $u(T) = v(T)$ then $u(t) = v(t)$ for all $t \in [0, T]$, i.e. that solutions have the 'backwards uniqueness property'. [Hint: assume that $\|w(0)\| \neq 0$ and show that $\|w(t)\| \neq 0$ for any $t \geq 0$. Take the inner product of (13.7) with w and divide by $L(t)\|w\|^2$; then use (13.9) and the fact that $\tilde{Q}(t)$ is bounded.]

14

The Takens Time-Delay Embedding Theorem

In his 1981 paper, 'Detecting strange attractors in turbulence', Takens showed that for a generic smooth system $x(t)$ evolving on a smooth d-dimensional manifold $\mathscr{M}$, the dynamics of solutions can be followed faithfully by taking k time-delayed copies of a 'generic measurement' $h : \mathscr{M} \to \mathbb{R}$

$$h(x), \quad h(x(T)), \quad \ldots, \quad h(x(kT)),$$

with $k \geq 2d$. More formally, he showed that for such an h, the mapping $\mathscr{M} \to \mathbb{R}^{k+1}$ given by

$$x \mapsto (h(x), h(x(T)), \ldots, h(x(kT)))$$

is a diffeomorphism.

Although the conclusions of his theorem are strong, so are its assumptions, which are hard to verify in general and may in fact fail in a number of practical applications. The requirement that the dynamics takes place on a compact finite-dimensional manifold is very restrictive, and excludes any direct application of the result to the attractors of infinite-dimensional dynamical systems. This means that in the form above this result provides no rigorous justification for the use of time-delay reconstruction for data from experiments in spatially extended systems.

In this chapter we show first how the result can be extended to the attractors of dynamical systems in $\mathbb{R}^N$ (following Sauer, Yorke, & Casdagli, (1991)), and then to the attractors of dynamical systems in infinite-dimensional spaces (following Robinson (2005)).

14.1 The finite-dimensional case

Sauer *et al.* (1991) replaced the manifold $\mathscr{M}$ with an invariant subset of $\mathbb{R}^N$ of (upper) box-counting dimension d and allowed dynamical systems that are only

Lipschitz continuous rather than smooth. In this section we give a generalised version of their argument that is valid for certain classes of Hölder continuous maps. We use this result in the next section to prove a version of the theorem valid for finite-dimensional invariant subsets of an infinite-dimensional space.

First we need to recall the definition of the singular values of a matrix and some of their properties. Let $M : \mathbb{R}^m \to \mathbb{R}^n$ be a linear map, and consider the $m \times m$ symmetric matrix $M^T M$. This matrix has a set of orthonormal eigenvectors $\{e_j\}_{j=1}^m$ with corresponding eigenvalues λ_j, i.e. $M^T M e_j = \lambda_j e_j$.

Lemma 14.1 *Each λ_j is nonnegative, and at most n of them are non-zero. The singular values of M are $\{\alpha_j\}_{j=1}^n$, where $\lambda_j = \alpha_j^2$. The vectors $\{Me_j\}_{j=1}^k$ corresponding to the nonzero values of λ_j are orthogonal in $\mathbb{R}^n$, with $|Me_j| = \alpha_j$.*

Proof Note that

$$\lambda_j = (\lambda_j e_j, e_j) = (M^T M e_j, e_j) = (Me_j, Me_j) = |Me_j|^2 \geq 0,$$

so that each eigenvalue is nonnegative. Next,

$$(Me_i, Me_j) = (M^T M e_i, e_j) = \lambda_j (e_i, e_j) = \lambda_j \delta_{ij},$$

so the $\{Me_i\}$ are orthogonal. Since there can be at most n mutually orthogonal vectors in $\mathbb{R}^n$, it follows that there are at most n nonzero eigenvalues of $M^T M$. □

We now use these singular values to describe the image of a ball in $\mathbb{R}^m$ under M. We write $B_m(r)$ for the ball in $\mathbb{R}^m$ of radius r, centred at the origin, and $B_m = B_m(1)$.

Lemma 14.2 *The image of the unit ball in $\mathbb{R}^m$ under a linear map M is an ellipse in $\mathbb{R}^n$ whose semiaxes are $\{Me_j\}_{j=1}^k$, where M has k non-zero singular values.*

Proof (After Section V 1.3 in Temam (1988).) We have already shown that the $\{Me_j\}$ are orthogonal in $\mathbb{R}^n$ and that $|Me_j| = \alpha_j$. Now take some $x \in B_m$; we can write

$$x = \sum_{j=1}^k x_j e_j + y \qquad \sum_{j=1}^k |x_j|^2 + |y|^2 \leq 1,$$

where y is orthogonal to the $\{x_j\}$. Then

$$Mx = \sum_{j=1}^k x_j (Me_j) = \sum_{j=1}^\infty (x_j \alpha_j) \frac{Me_j}{\alpha_j}.$$

This expresses Mx as $\sum_j \xi_j \hat{e}_j$ where the $\{\hat{e}_j\}$ are orthonormal vectors in the directions of Me_j; clearly

$$\sum_j \left(\frac{\xi_j}{\alpha_j}\right)^2 \leq 1,$$

and so the image MB_m is an ellipse as stated. □

We now prove a simple lemma (after Lemma 4.2 of Sauer *et al.* (1991)), which provides the 'key inequality' for the finite-dimensional time-delay theorem, just as Lemma 4.1 provided the key inequality for the finite-dimensional Hölder embedding result of Theorem 4.3.

Lemma 14.3 *Let $M : \mathbb{R}^m \to \mathbb{R}^n$ be a linear map. For a positive integer r ($1 \leq r \leq n$), let $\alpha_r > 0$ be the rth largest singular value of M. Then for any $b \in \mathbb{R}^n$,*

$$\frac{\text{Vol}\,\{x \in B_m(\rho) : |Mx + b| < \delta\}}{\text{Vol}(B_m(\rho))} \leq C_{m,n} \left(\frac{\delta}{\alpha_r \rho}\right)^r. \tag{14.1}$$

Proof We have just shown that the image of $B_m(\rho)$ under M, $MB_m(\rho)$, is an ellipse, whose semiaxes are $\{\rho\alpha_i\}$, where the α_i are the singular values of M (Lemma 14.2). It follows that decreasing the size of any of the singular values of M can only shrink the size of the image, and so increase the value of the left-hand side of (14.1). So we can assume that $\alpha_1 = \cdots = \alpha_r > 0$, and $\alpha_{r+1} = \cdots = \alpha_n = 0$, which means that $MB_m(\rho)$ is an r-dimensional ball of radius $\rho\alpha_r$. It is clear that the intersection $MB_m(\rho) + b$ with $B_n(\delta)$ is maximised when $b = 0$, and hence

$$\text{Vol}\{x \in B_m(\rho) : |Mx + b| < \delta\} \leq \Omega_r \Omega_{m-r} \left(\frac{\delta}{\alpha_r}\right)^r \rho^{m-r}.$$

Since $\text{Vol}\,B_m(\rho) = \Omega_m \rho^m$ and $1 \leq r \leq n$, the left-hand side of (14.1) is bounded by

$$\left(\max_{1 \leq r \leq n} \frac{\Omega_r \Omega_{m-r}}{\Omega_m}\right) \left(\frac{\delta}{\alpha_r \rho}\right)^r \leq C_{m,n} \left(\frac{\delta}{\alpha_r \rho}\right)^r.$$ □

We now use this bound to prove the following lemma, which is the main component of the proof of the Takens time-delay theorem; it is a version of Lemma 4.6 from Sauer *et al.* (1991), but valid for Hölder continuous maps $\{F_j\}$. In some sense it is a generalised version of Theorem 4.3, but without the Hölder continuity of the inverse; the embedding part of that theorem is an immediate corollary if one takes $F_0, \ldots, F_m$ to be a basis for the linear maps from $\mathbb{R}^k$ into $\mathbb{R}^n$ (so that $\theta = 1$), and notes that in this case $M_{(x,y)}$ always has rank k (see Exercise 14.2).

Lemma 14.4 *Let A be a compact subset of $\mathbb{R}^N$, and let $\{F_0, F_1, \ldots, F_m\}$ be θ-Hölder maps from A into $\mathbb{R}^k$. Let S_r be the set of pairs $x \neq y$ in A for which the $k \times m$ matrix*

$$M_{(x,y)} = (F_1(x) - F_1(y) \qquad \cdots \qquad F_m(x) - F_m(y))$$

has rank at least r, and suppose that $S_0 = \emptyset$ and $d_{\mathrm{B}}(\overline{S_r}) < r\theta$ for $1 \leq r \leq k$. Then for almost every $\alpha \in \mathbb{R}^m$ the map $F_\alpha : X \to \mathbb{R}^k$ given by

$$F_\alpha = F_0 + \sum_{j=1}^{m} \alpha_j F_j$$

is one-to-one on A.

We follow Sauer *et al.* (1991) and say that 'G_α has property Γ with probability p' if the Lebesgue measure of the set of $\alpha \in B_m(R)$ for which G_α has property Γ is p times the measure of $B_m(R)$.

Proof For $j = 0, \ldots, m$ set $G_j(x, y) = F_j(x) - F_j(y)$, and let G_α be the map given by

$$G_\alpha(x, y) = G_0(x, y) + \sum_{j=1}^{m} \alpha_j G_j(x, y),$$

so that by assumption for each $z = (x, y) \in S_r$ the $k \times m$ matrix

$$M_z = (G_1(z) \qquad \cdots \qquad G_m(z))$$

has rank at least r.

Fix $R > 0$, and consider the set

$$S_{r,j} = \{z \in S_r : r\text{th largest singular value of } M_z > 1/j\}.$$

Since the rank of M_z is at least r for every $z \in S_r$, the rth largest singular value is always positive, and so $S_r = \cup_{j=1}^{\infty} S_{r,j}$.

Note that

$$G_\alpha(z) = G_0(z) + M_z\alpha,$$

and so Lemma 14.3 implies that for $z \in S_{r,j}$ the probability that $|G_\alpha(z)| < \delta$ is no larger than $C_{m,n}(j\delta/R)^r$.

Now fix r and j, and choose d with $d_{\mathrm{B}}(\overline{S_r}) < d < r\theta$. Since $d > d_{\mathrm{B}}(\overline{S_r})$ there exists an $\epsilon_0 > 0$ such for any $0 < \epsilon < \epsilon_0$, $\overline{S}_{r,j}$ can be covered by no more than ϵ^{-d} balls of radius ϵ, $\{B(z_k, \epsilon)\}$. Set

$$Y_k = S_{r,j} \cap B(z_k, \epsilon).$$

Since $|G_\alpha(x) - G_\alpha(y)| \leq K|x - y|^\theta$,

$$|G_\alpha(z_k)| > K\epsilon^\theta \qquad \Rightarrow \qquad |G_\alpha(z)| > 0 \quad \text{for all} \quad z \in Y_k;$$

so to have $G_\alpha(z) = 0$ for some $z \in Y_k$ requires $|G_\alpha(z_k)| \leq K\epsilon^\theta$.

For each fixed z_k, Lemma 14.3 guarantees that the probability that

$$|G_\alpha(z_k)| = |G_0(z_k) + M_{z_k}\alpha| \leq K\epsilon^\theta$$

is at most $C_{t,n}(jK\epsilon^\theta/R)^r$. Since there are no more than ϵ^{-d} of the $\{z_k\}$, it follows that the probability that $G_\alpha(z) = 0$ for some $z \in S_{r,j}$ is bounded by

$$\epsilon^{-d} \times C_{m,n}(jK/R)^r\epsilon^{\theta r} = C_{m,n,j,R,r}\epsilon^{\theta r - d}.$$

Since $d < \theta r$ and ϵ is arbitrary, it follows that $G_\alpha(z) \neq 0$ for all $z \in S_{r,j}$ with probability one.

Thus $G_\alpha(z) \neq 0$ for all $z \in S_r$ with probability 1, and since $A = \cup_{r=1}^k S_r$, it follows that $G_\alpha(z) \neq 0$ for all $z \in A$ with probability 1, i.e. for almost every $\alpha \in B_m(R)$. Since $R > 0$ was arbitrary, $G_\alpha(z) \neq 0$ for all $z \in A$ for almost every $\alpha \in \mathbb{R}^m$. □

We now prove a finite-dimensional version of the Takens Theorem, after Theorem 4.13 of Sauer *et al.* (1991). However, this version works for certain Hölder continuous maps, namely maps g for which all iterates g^r have the same Hölder exponent. This condition that g and its iterates have the same Hölder exponent does not appear a natural one (in general, if g is θ-Hölder then g^p will be θ^p-Hölder). However, it is satisfied automatically if g is Lipschitz ($\theta = 1$), and will in particular be satisfied by the class of maps we consider in the next section, namely maps of the form

$$g(x) = L^{-1} \circ \Phi \circ L(x),$$

defined on $X = L\mathscr{A}$, where $L : \mathscr{B} \to \mathbb{R}^k$ is linear, $\Phi : \mathscr{B} \to \mathscr{B}$ is Lipschitz, and $L^{-1} : X \to \mathscr{A}$ is θ-Hölder, since for such maps

$$g^p(x) = L^{-1} \circ \Phi^p \circ L(x).$$

The statement of the theorem makes explicit the sense in which the set of Hölder functions $h : \mathbb{R}^N \to \mathbb{R}$ such that $F_k[h, g]$ is one-to-one on X is 'prevalent'.

Theorem 14.5 *Let X be a compact subset of $\mathbb{R}^N$ with $d_B(X) = d$, and let $g : X \to X$ be such that g^r is a θ-Hölder function for any $r \in \mathbb{N}$. Take $k > 2d/\theta$*

($k \in \mathbb{N}$), and assume that the set X_p of p-periodic points of g (i.e. $x \in X$ such that $g^p(x) = x$) satisfies $d_B(X_p) < p\theta^2/2$ for all $p = 1, \ldots, k$.

Let $h_1, \ldots, h_m$ be a basis for the polynomials in N variables of degree at most 2k, and given any θ-Hölder function $h_0 : \mathbb{R}^N \to \mathbb{R}$ define

$$h_\alpha = h_0 + \sum_{j=1}^{m} \alpha_j h_j.$$

Then for almost every $\alpha \in \mathbb{R}^m$ the k-fold observation map $F_k : X \to \mathbb{R}^N$ defined by

$$F_k[h_\alpha, g](x) = \left(h_\alpha(x), h_\alpha(g(x)), \ldots, h_\alpha(g^{k-1}(x))\right)^T$$

is one-to-one on X.

(The condition that iterates of g be θ-Hölder is in fact only required for $g, \ldots, g^k$.)

Proof For $i = 0, 1, \ldots, m$ define

$$F_i(x) = \begin{pmatrix} h_i(x) \\ h_i(g(x)) \\ \vdots \\ h_i(g^{k-1}(x)) \end{pmatrix},$$

so that by definition

$$F_k(h_\alpha, g) = F_0 + \sum_{j=1}^{m} \alpha_j F_j.$$

In order to apply Lemma 14.4 we need to check, for each $x \neq y$, the rank of the matrix

$$\begin{aligned} M_{(x,y)} &= (F_1(x) - F_1(y) \quad \cdots \quad F_m(x) - F_m(y)) \\ &= \begin{pmatrix} h_1(x) - h_1(y) & \cdots & h_m(x) - h_m(y) \\ \vdots & \ddots & \vdots \\ h_1(g^{k-1}(x)) - h_1(g^{k-1}(y)) & \cdots & h_m(g^{k-1}(x)) - h_m(g^{k-1}(y)) \end{pmatrix}. \end{aligned}$$

In order to analyse this, it is helpful to write M in the form $M = JH$, where

$$H_{(x,y)} = \begin{pmatrix} h_1(z_1) & \cdots & h_m(z_1) \\ \vdots & \ddots & \vdots \\ h_1(z_q) & \cdots & h_m(z_q) \end{pmatrix},$$

with all of the $z_1, \ldots, z_q$ distinct (we have $q \leq 2k$), and where $J_{(x,y)}$ is a $k \times q$ matrix each of whose rows consists of zeros except for one 1 and one -1. Given any $\xi \in \mathbb{R}^q$, we can find[1] a set of coefficients $\{\alpha_j\}_{j=1}^m$ such that $\sum_j \alpha_j h_j(z_l) = \xi_l$, i.e. such that $H_{(x,y)}\alpha = \xi$. This implies that the rank of H is q; since $J : \mathbb{R}^q \to \mathbb{R}^k$, we only need to check the rank of J.

We split the set $\{(x, y) : x, y \in X, x \neq y\}$ into three disjoint sets of pairs (x, y), and show that $F_\alpha(x) \neq F_\alpha(y)$ for almost every α on each of these sets. It then follows that $F_\alpha(x) \neq F_\alpha(y)$ for almost every α, for any $x, y \in X$ with $x \neq y$.

Case 1: x and y are not both periodic of period $\leq k$. In this case without loss of generality $\{x, g(x), \ldots, g^{k-1}(x)\}$ consists of k discrete points and $\{y, \ldots, g^{k-1}(y)\}$ consists of $r \geq 1$ points distinct from the iterates of x. So

$$J_{(x,y)} = \left(\begin{array}{cccccc|ccccc} 1 & 0 & \cdots & \cdots & 0 & & -1 & 0 & \cdots & \cdots & 0 \\ 0 & 1 & \cdots & \cdots & 0 & & 0 & -1 & \cdots & \cdots & 0 \\ 0 & 0 & \ddots & \cdots & 0 & & 0 & 0 & \vdots & \cdots & 0 \\ \vdots & \vdots & \vdots & \ddots & 0 & & \vdots & \vdots & \vdots & \vdots & \vdots \\ 0 & 0 & \cdots & 0 & 1 & & 0 & \cdots & -1 & \cdots & 0 \end{array}\right),$$

where the left block is $k \times k$ and the right block is $r \times k$ (with the top $r \times r$ entries being minus the identity).

It follows that the rank of $J : \mathbb{R}^{k+r} \to \mathbb{R}^k$ is k, and so is the rank of $F = JH$. Since the set of pairs $x \neq y$ has box-counting dimension at most $2d$, and we have just shown that rank $M_{(x,y)} = k > 2d/\theta$ by assumption, the conditions of Lemma 14.4 are met for this choice of (x, y), i.e. for all such (x, y), $F_\alpha(x) \neq F_\alpha(y)$ for almost every α.

Case 2: x and y lie in distinct periodic orbits of period $\leq k$. Suppose that p and q are the minimal integers such that $g^p(x) = x$ and $g^q(y) = y$, without loss of generality $1 \leq q \leq p \leq k$. Then J has rank at least p (its top left $p \times p$ entries are the $p \times p$ identity matrix). So rank $M_{(x,y)} \geq p$ in this case, while by assumption the set of pairs of periodic points of period $\leq p$ has dimension $< p/\theta$. So once more we can apply Lemma 14.4.

Case 3: x and y lie on the same periodic orbit of period $\leq k$. Suppose that p and q are the minimal integers such that $g^p(x) = x$ and $g^q(x) = y$, with $1 \leq q < p \leq k$. As an illustrative example, if $p = 7$ and $q = 4$, then J is of

[1] One can make a linear change of coordinates so that the first components of $z_1, \ldots, z_q$ are distinct, and then simply interpolate in one variable.

the form

$$J_{(x,y)} = \begin{pmatrix} 1 & 0 & 0 & 0 & -1 & 0 & 0 \\ 0 & 1 & 0 & 0 & 0 & -1 & 0 \\ 0 & 0 & 1 & 0 & 0 & 0 & -1 \\ -1 & 0 & 0 & 1 & 0 & 0 & 0 \\ 0 & -1 & 0 & 0 & 1 & 0 & 0 \\ 0 & 0 & -1 & 0 & 0 & 1 & 0 \\ 0 & 0 & 0 & -1 & 0 & 0 & 1 \end{pmatrix}.$$

The rank of such a matrix is at least $p/2$, and hence rank $M_{(x,y)} \geq p/2$ in this case. The box-counting dimension of the set of pairs that lie on the same periodic orbit is bounded by $d_{\mathrm{B}}(A_p)/\theta$, since any such (x, y) is contained in the image of A_p under one of the mappings $x \mapsto (x, g^j(x))$ for some $j = 1, \ldots, k$. □

14.2 Periodic orbits and the Lipschitz constant for ordinary differential equations

If we recast Theorem 14.5 in terms of a continuous flow generated by a Lipschitz ordinary differential equation (so $g = S(T)$ for some $T > 0$ and $\theta = 1$) then observe that there can be no embedding result if there are periodic orbits of period T or even $2T$. This follows from the statement of the theorem (since in this case the dimension of X_1 and X_2 is at least 1, and the condition $\dim(X_p) < p/2$ cannot be satisfied for $p = 1, 2$), but this is not simply an artefact of the proof, as the following arguments from Sauer *et al.* (1991) show.

If there is a periodic orbit Γ of period T this is a topological circle. But under the time-delay mapping, any point on Γ maps onto a line in $\mathbb{R}^k$. One cannot map a circle onto a line using any continuous one-to-one mapping, so the theorem must fail in this case.

Such an embedding is also impossible if there is a periodic orbit Γ of period 2. Consider the map $x \mapsto \tilde{h}(x) = h(g(x)) - h(x)$. Then either $\tilde{h}(x) \equiv 0$ on Γ, or there is some $x_0 \in \Gamma$ such that $\tilde{h}(x_0) \neq 0$. In the latter case,

$$\tilde{h}(g(x_0)) = h(g^2(x_0)) - h(g(x_0)) = h(x_0) - h(g(x_0)) = -\tilde{h}(x_0),$$

so that there must be some $x^* \in \Gamma$ with $\tilde{h}(x^*) = 0$. But this implies that $h(x^*) = h(g(x^*))$, so that x^* and $g(x^*)$ (which are distinct points) are mapped to the same point in $\mathbb{R}^k$ whatever the value of k.

So it is interesting if we can find a T sufficiently small that there are no periodic orbits of any period $\leq T$. Yorke (1969) showed that if the ordinary differential equation

$$\dot{x} = f(x) \qquad \text{with} \qquad |f(x) - f(y)| \le L|x - y|$$

has a periodic orbit of period T, then one must have $T \ge 2\pi/L$. (Yorke also gives an example to show that the factor of 2π is sharp.) We give a version of this result here, following ideas in Kukavica (1994); the result is no longer sharp (we only obtain $T \ge 1/L$), but this sacrifice seems worthwhile given the simplicity of the proof, which also provides a model for the proof of a similar infinite-dimensional result (Theorem 14.8).

Theorem 14.6 *Any periodic orbit of the equation $\dot{x} = f(x)$, where f has Lipschitz constant L, has period $T \ge 1/L$.*

Proof Fix $\tau > 0$ and set $v(t) = x(t) - x(t - \tau)$. Then

$$v(t) - v(s) = \int_s^t \dot{v}(r)\,\mathrm{d}r.$$

Integrating both sides with respect to s from 0 to T gives

$$T v(t) = \int_0^T \left(\int_s^t \dot{v}(r)\,\mathrm{d}r \right) \mathrm{d}s$$

since $\int_0^T v(s)\,\mathrm{d}s = 0$ because x is T-periodic. Thus

$$T|v(t)| \le \int_0^T \int_0^T |\dot{v}(r)|\,\mathrm{d}r\,\mathrm{d}s \le T \int_0^T |\dot{v}(r)|\,\mathrm{d}r,$$

i.e.

$$\begin{aligned} |x(t) - x(t - \tau)| \le \int_0^T |\dot{v}(s)|\,\mathrm{d}s &= \int_0^T |f(x(s)) - f(x(s - \tau))|\,\mathrm{d}s \\ &\le L \int_0^T |x(s) - x(s - \tau)|\,\mathrm{d}s. \end{aligned}$$

Therefore

$$\int_0^T |x(t) - x(t - \tau)|\,\mathrm{d}t \le LT \int_0^T |x(s) - x(s - \tau)|\,\mathrm{d}s,$$

and it follows that if $LT < 1$ then

$$\int_0^T |x(t) - x(t - \tau)|\,\mathrm{d}t = 0.$$

Thus $x(t) = x(t - \tau)$ for all $\tau > 0$, i.e. $x(t)$ is constant. □

14.3 The infinite-dimensional case

We now prove an infinite-dimensional version of the Takens Theorem: we use the Hölder embedding theorem for sets with finite box-counting dimension (Theorem 8.1) to produce a finite-dimensional system to which we can apply the finite-dimensional result of Theorem 14.5.

Theorem 14.7 *Let $\mathscr{A}$ be a compact subset of a Banach space $\mathscr{B}$ with upper box-counting dimension $d_{\rm B}(\mathscr{A}) = d$ and dual thickness $\tau^*(\mathscr{A}) = \tau$. Set $\alpha = 1/2$ if $\mathscr{B}$ is a Hilbert space; otherwise take $\alpha = 1$.*

Suppose that $\mathscr{A}$ is an invariant set for a Lipschitz map $\Phi : \mathscr{B} \to \mathscr{B}$; choose an integer $k > 2(1 + \alpha\tau)d$, and suppose further that the set $\mathscr{A}_p$ of p-periodic points of Φ satisfies $(1 + \alpha\tau)^2 d_{\rm B}(\mathscr{A}_p) < p/2$ for $p = 1, \ldots, k - 1$. Then a prevalent set of Lipschitz maps $f : \mathscr{B} \to \mathbb{R}$ make the k-fold observation map $D_k[f, \Phi] : \mathscr{B} \to \mathbb{R}^k$ defined by

$$D_k[f, \Phi](u) = \big(f(u), f(\Phi(u)), \ldots, f(\Phi^{k-1}(u))\big)$$

one-to-one on $\mathscr{A}$.

Proof Given $k > 2(1 + \alpha\tau)d$, first choose N large enough that

$$k > \frac{2N(1 + \alpha\tau)}{N - 2d} d \qquad \text{and} \qquad d_{\rm B}(\mathscr{A}_p) < \left[\frac{N - 2d}{N(1 + \alpha\tau)}\right]^2 \frac{p}{2}$$

for $p = 1, \ldots, k$, and then pick $\alpha < (N - 2d)/[N(1 + \alpha\tau)]$ such that $k > 2d/\alpha$ and $d_{\rm B}(\mathscr{A}_p) > \alpha^2 p/2$ for $p = 1, \ldots, k$.

Use Theorem 8.1 to find a bounded linear function $L : \mathscr{B} \to \mathbb{R}^N$ that is one-to-one on $\mathscr{A}$ and satisfies

$$\|x - y\| \le c|Lx - Ly|^\alpha \qquad \text{for all} \qquad x, y \in \mathscr{A}.$$

The set $X = L\mathscr{A} \subset \mathbb{R}^N$ is an invariant set for the induced mapping $g : X \to X$ defined by

$$g(\xi) = L\Phi(L^{-1}\xi).$$

Since

$$g^n(\xi) = L\Phi^n(L^{-1}\xi)$$

all the iterates of g are α-Hölder:

$$\begin{aligned} |g^n(\xi) - g^n(\eta)| &= |L\Phi^n(L^{-1}\xi) - L\Phi^n(L^{-1}\eta)| \\ &\le \|L\||\Phi^n(L^{-1}\xi) - \Phi^n(L^{-1}\eta)| \\ &\le l_\Phi^n \|L\||L^{-1}\xi - L^{-1}\eta| \\ &\le c\, l_\Phi^n \|L\||\xi - \eta|^\alpha, \end{aligned}$$

where $\|L\|$ is the operator norm of $L : \mathscr{B} \to \mathbb{R}^N$ and l_Φ is the Lipschitz constant of $\Phi : \mathscr{B} \to \mathscr{B}$.

Observe that if x is a fixed point of Φ^j then $\xi = Lx$ is a fixed point of g^j, and vice versa. It follows that X_p, the set of all points of X that are p-periodic for g, is given simply by $X_p = L\mathscr{A}_p$. Since L is Lipschitz and the box-counting dimension does not increase under the action of Lipschitz maps (Lemma 3.3(iv)), $d_{\rm B}(X_p) = d_{\rm B}(L\mathscr{A}_p) \leq d_{\rm B}(\mathscr{A}_p) < p/2(1+\alpha\tau)^2$. Similarly, $d_{\rm B}(X) \leq d_{\rm B}(\mathscr{A})$.

Given a Lipschitz map $f_0 : \mathscr{A} \to \mathbb{R}$, define the α-Hölder map $h_0 : X \to \mathbb{R}$ by

$$h_0(\xi) = f_0(L^{-1}\xi) \qquad \text{for all} \qquad \xi \in X.$$

With $\{h_j\}_{j=1}^m$ a basis for the polynomials in N variables of degree at most $2k$, all the conditions of Theorem 14.5 are satisfied, and hence for almost every $\alpha \in \mathbb{R}^m$, the k-fold observation map on $\mathbb{R}^N$ given by

$$F_k[h_\alpha, g](\xi) = \left(h_\alpha(\xi), h_\alpha(g(\xi)), \ldots, h_\alpha(g^{k-1}(\xi))\right)^T,$$

where

$$h_\alpha(x) = h_0(x) + \sum_{j=1}^m \alpha_j h_j(x),$$

is one-to-one on X.

Now consider the k-fold observation map on $\mathscr{A}$ given by

$$F_k[h_\alpha, g](Lx) = \left(h_\alpha(Lx), h_\alpha(L\Phi(x)), \cdots, h_\alpha(L\Phi^{k-1}(x))\right)^T.$$

Since L is one-to-one between $\mathscr{A}$ and X, and $F_k[h, \alpha, g]$ is one-to-one between X and its image, it follows that $F_k \circ L$ is one-to-one between $\mathscr{A}$ and its image.

If we define $f_j(x) = h_j(Lx)$, then each f_j is a Lipschitz map from $\mathscr{A}$ into $\mathbb{R}^k$, and we can write

$$F_k[h_\alpha, g](Lx) = D_k[f_\alpha, \Phi](x) = \left(f_\alpha(x), f_\alpha(\Phi(x)), \ldots, f_\alpha(\Phi^{k-1}(x))\right)^T,$$

where

$$f_\alpha = f_0 + \sum_{j=1}^M \alpha_j f_j.$$

Then if we take

$$E = \left\{\sum_{j=1}^M \alpha_j f_j : (\alpha_1, \ldots, \alpha_M) \in B_M(0, 1)\right\}$$

and equip E with the measure induced by the uniform measure on $B_M(0, 1)$, it follows that a prevalent set of Lipschitz $f : \mathscr{A} \to \mathbb{R}^k$ make the map $D_k[f, \Phi]$ one-to-one on $\mathscr{A}$. □

Note that the condition on the number of delay coordinates required increases with the dual thickness of the set $\mathscr{A}$. In the case when $\mathscr{A}$ has zero dual thickness (so, for example, if $\mathscr{A}$ is 'smooth' so that Lemma 13.1 guarantees that its thickness is zero, or if the equation is in the right form that Theorem 13.3 shows that the Lipschitz deviation of $\mathscr{A}$ is zero) then the condition on k reduces to the $k > 2d$ one would obtain in the finite-dimensional Lipschitz case (Theorem 14.5 with $\theta = 1$).

14.4 Periodic orbits and the Lipschitz constant for semilinear parabolic equations

As before, when $\Phi = S(T)$ (the time T map of some underlying continuous time flow) the condition $d_{\mathrm{B}}(\mathscr{A}_p) < p/2(1 + \alpha\tau)$ precludes the existence of periodic orbits of periods pT for all integers p such that $p \leq 2 + \alpha\tau$. The following infinite-dimensional generalisation of Theorem 14.6, due to Robinson & Vidal-López (2006), is therefore useful.

The result treats the abstract semilinear parabolic equation

$$\mathrm{d}u/\mathrm{d}t = -Au + g(u) \tag{14.2}$$

as considered in Section 13.2; the argument relies crucially on the fact that the solution $u(t)$ is given by the variation of constants formula

$$u(t) = \mathrm{e}^{-At}u_0 + \int_0^t \mathrm{e}^{-A(t-s)}g(u(s))\,\mathrm{d}s.$$

Theorem 14.8 *For each α with $0 \leq \alpha \leq 1/2$ there exists a constant K_α such that if*

$$\|g(u) - g(v)\| \leq L\|A^\alpha(u - v)\| \qquad \textit{for all} \qquad u, v \in D(A^\alpha)$$

then any periodic orbit of (14.2) must have period at least $K_\alpha L^{-1/(1-\alpha)}$.

While we assume that g is uniformly Lipschitz, this uniformity need only hold for u, v contained in the (necessarily bounded) periodic orbit.

Proof On a periodic orbit of period T we have

$$u(t) = u(t + T) = \mathrm{e}^{-AT}u(t) + \int_0^T \mathrm{e}^{-A(T-s)}g(u(s + t))\,\mathrm{d}s,$$

and so

$$(I - \mathrm{e}^{-AT})u(t) = \int_0^T \mathrm{e}^{-A(T-s)} g(u(s+t))\,\mathrm{d}s.$$

It follows that

$$\begin{aligned} &u(t) - u(t+\tau) \\ &\quad = (I - \mathrm{e}^{-AT})^{-1} \int_0^T \mathrm{e}^{-A(T-s)}[g(u(t+s)) - g(u(t+\tau+s))]\,\mathrm{d}s. \end{aligned}$$

Since u is T-periodic,

$$\int_0^T g(u(s+t))\,\mathrm{d}s = \int_0^T g(u(s+t+\tau))\,\mathrm{d}s,$$

and so in fact for any constant c

$$\begin{aligned} &(I - \mathrm{e}^{-AT})(u(t) - u(t+\tau)) \\ &\qquad = \int_0^T (\mathrm{e}^{-A(T-s)} - cI)(g(u(s+t)) - g(u(s+t+\tau)))\,\mathrm{d}s. \end{aligned}$$

Therefore

$$\begin{aligned} &u(t) - u(t+\tau) \\ &\quad = \int_0^T \left[(I - \mathrm{e}^{-AT})^{-1}(\mathrm{e}^{-A(T-s)} - cI)\right](g(u(s+t)) - g(u(s+t+\tau)))\,\mathrm{d}s. \end{aligned}$$

For ease of notation we now write

$$D(t) = u(t) - u(t+\tau) \qquad \text{and} \qquad G(t) = g(u(t)) - g(u(t+\tau)).$$

Then since the eigenfunctions of A are also the eigenfunctions of

$$(I - \mathrm{e}^{-AT})^{-1}(\mathrm{e}^{-A(T-s)} - cI),$$

we have, for each $k \in \mathbb{N}$,

$$(A^\alpha D(t), w_k) = \int_0^T \lambda_k^\alpha \frac{\mathrm{e}^{-\lambda_k(T-s)} - c}{1 - \mathrm{e}^{-\lambda_k T}} (G(t+s), w_k)\,\mathrm{d}s,$$

and so

$$\begin{aligned} &|(A^\alpha D(t), w_k)| \\ &\quad \le \frac{\lambda_k^\alpha}{1 - \mathrm{e}^{-\lambda_k T}} \left(\int_0^T (\mathrm{e}^{-\lambda_k s} - c)^2\,\mathrm{d}s\right)^{1/2} \left(\int_0^T (G(t+s), w_k)^2\,\mathrm{d}s\right)^{1/2}. \end{aligned}$$

We now choose $c = (1 - \mathrm{e}^{-\lambda_k T})/\lambda_k T$ in order to minimise the first integral, for which we then obtain

$$\int_0^T (\mathrm{e}^{-\lambda_k s} - c)^2 \, \mathrm{d}s = T \left[\frac{1 - \mathrm{e}^{-2\lambda_k T}}{2\lambda_k T} - \frac{(1 - \mathrm{e}^{-\lambda_k T})^2}{(\lambda_k T)^2} \right].$$

Therefore

$$|(A^\alpha D(t), w_k)| \le T^{1/2-\alpha} \, \Phi(\lambda_k T) \left(\int_0^T (G(t+s), w_k)^2 \, \mathrm{d}s \right)^{1/2},$$

where

$$\Phi(\mu) := \frac{\mu^\alpha}{1 - \mathrm{e}^{-\mu}} \left[\frac{1 - \mathrm{e}^{-2\mu}}{2\mu} - \frac{(1 - \mathrm{e}^{-\mu})^2}{\mu^2} \right]^{1/2}.$$

Now, $\Phi(\mu)$ is bounded on $[0, \infty)$ by some constant C_α: it is clear that $\Phi(\mu) \sim \mu^{\alpha - 1/2}/\sqrt{2}$ as $\mu \to \infty$, while a careful Taylor expansion shows that $\Phi(\mu) \sim \mu^\alpha / 2\sqrt{3}$ as $\mu \to 0$ (see Exercise 14.4).

It follows that for each $k \in \mathbb{N}$

$$|(A^\alpha D(t), w_k)|^2 \le C_\alpha^2 \, T^{1-2\alpha} \int_0^T |(G(t+s), w_k)|^2 \, \mathrm{d}s.$$

Summing both sides over all k we obtain

$$|A^\alpha D(t)|^2 \le C_\alpha^2 \, T^{1-2\alpha} \int_0^T |G(t+s)|^2 \, \mathrm{d}s \le C_\alpha^2 \, T^{1-2\alpha} L^2 \int_0^T |A^\alpha D(s)|^2 \, \mathrm{d}s.$$

Now integrate the left- and right-hand sides of this expression with respect to t between $t = 0$ and $t = T$ to obtain

$$\int_0^T |A^\alpha D(t)|^2 \, \mathrm{d}t \le C_\alpha^2 \, T^{2-2\alpha} L^2 \int_0^T |A^\alpha D(s)|^2 \, \mathrm{d}s.$$

Therefore if $C_\alpha T^{1-\alpha} L < 1$ we must have

$$\int_0^T |A^\alpha (u(t) - u(t+\tau))|^2 \, \mathrm{d}t = 0.$$

It follows that $u(t) = u(t+\tau)$ for all $t \in [0, T]$, and since this holds for any $\tau > 0$, $u(t)$ must be a constant orbit. Therefore any periodic orbit must have period at least $K_\alpha L^{-1/(1-\alpha)}$. □

Exercises

14.1 Show that the nonzero singular values of M are also the square roots of the eigenvalues of MM^T.

14.2 Choose some $L \in \mathscr{L}(\mathbb{R}^N, \mathbb{R}^k)$, and let $\{e_\alpha\}_{\alpha=1}^N$ be a basis for $\mathbb{R}^N$, and $\{\hat{e}_j\}_{j=1}^k$ be a basis for $\mathbb{R}^k$. Write L in the form

$$L = \sum_{\alpha,j} c_{\alpha,j} L_{\alpha,j},$$

where $L_{\alpha,j}$ is the linear map from $\mathbb{R}^N$ into $\mathbb{R}^k$ given by

$$L_{\alpha,j}(z) = (z, e_\alpha)\hat{e}_j \qquad \alpha = 1, \ldots, N; \; j = 1, \ldots, k$$

(i.e. $L_{\alpha,j}$ is the linear map that sends e_α to $\hat{e}_j$). Write $Lz = M_z c$ for some transformation M_z from $\mathbb{R}^{Nk}$ into $\mathbb{R}^k$ and $c \in \mathbb{R}^{Nk}$, and show that M_z has k nonzero singular values, all of which are $|z|$. [Hint: use the result of the previous exercise to find the singular values of M_z.]

14.3 Let E be the set of linear maps $L : \mathbb{R}^N \to \mathbb{R}^k$ of the form

$$L = (l_1^*, l_2^*, \cdots, l_N^*),$$

with $\sum_{\alpha=1}^N |l_\alpha|^2 \leq 1$; this is equivalent to taking $\{l_{\alpha,j}\}$ in B_{Nk}, the unit ball in $\mathbb{R}^{Nk}$. Equip E with a probability measure μ, equal to Lebesgue measure on B_{Nk}, normalised so that the total measure of E is equal to one. Use Lemma 14.3 combined with the result of the previous exercise to show that for any $x \in \mathbb{R}^N$ and any $\epsilon > 0$,

$$\mu\{L \in E : |Lx| < \epsilon\} \leq c \left(\frac{\epsilon}{|x|} \right)^k,$$

where c depends on k and N (cf. Lemma 4.1).

14.4 Perform a Taylor expansion of

$$\Phi(\mu) = \frac{\mu^\alpha}{1 - \mathrm{e}^{-\mu}} \left[\frac{1 - \mathrm{e}^{-2\mu}}{2\mu} - \frac{(1 - \mathrm{e}^{-\mu})^2}{\mu^2} \right]^{1/2}.$$

about $\mu = 0$ to show that $\Phi(\mu) \sim \mu^\alpha / 2\sqrt{3}$ as $\mu \to 0$.

15

Parametrisation of attractors via point values

The aim of this final chapter is to show that if the attractor $\mathscr{A}$ consists of real analytic functions defined on some domain Ω then one can parametrise the attractor using a sufficient number of point values. We will show that if k is large enough (proportional to the box-counting dimension of $\mathscr{A}$) then for almost every choice of k points $\{x_j\}$ in Ω the mapping

$$u \mapsto (u(x_1), \ldots, u(x_k))$$

is an embedding of $\mathscr{A}$ into $\mathbb{R}^{kd}$.

More precisely, we will give a proof of the following theorem, first proved (in a slightly different form) by Friz & Robinson (2001). The proof makes use of a number of the ideas that have already been discussed: both the Hausdorff and box-counting dimensions, the thickness and dual thickness, and, of course, the embedding result of Theorem 8.1. (For a related result in the context of purely analytic systems see Sontag (2002).)

Theorem 15.1 *Let $\mathscr{A}$ be a compact subset of $L^2_{\rm per}(\Omega, \mathbb{R}^d)$ with $d_{\rm B}(\mathscr{A})$ finite. Suppose also that $\mathscr{A}$ consists of real analytic functions. Then, for $k \geq 16 d_{\rm B}(\mathscr{A}) + 1$ almost every set $\mathbf{x} = (x_1, \ldots, x_k)$ of k points in Ω makes the map $E_{\mathbf{x}}$, defined by*

$$E_{\mathbf{x}}[u] = (u(x_1), \ldots, u(x_k))$$

one-to-one between X and its image.

In the statement of the theorem, we set $\Omega = \prod_{j=1}^{m} [0, L_j]$, and denote by $L^2_{\rm per}(\Omega, \mathbb{R}^d)$ those functions in $L^2_{\rm loc}(\mathbb{R}^m, \mathbb{R}^d)$ that are periodic with period $L_j > 0$ in the $\{e_j\}$ direction,

$$u(x + L_j e_j) = u(x) \qquad \text{for all} \qquad j = 1, \ldots, m.$$

'Almost every' is with respect to Lebesgue measure on Ω^k.

Although we give the result for attractors that consist of periodic functions, with a little additional work essentially the same techniques can be used to prove a more general result, valid (for example) on bounded domains with Dirichlet boundary conditions, see Kukavica & Robinson (2004).

Before starting the proof proper, we give an idea of why the analyticity condition is required. Suppose that we have chosen $(x_1, \ldots, x_k)$, and that these are in fact a 'bad' set of points, which means that there are $u, v \in \mathscr{A}$ with $u \neq v$ such that

$$u(x_j) = v(x_j) \qquad \text{for every} \quad j = 1, \ldots, k.$$

Looking instead at the set of differences $X = \mathscr{A} - \mathscr{A}$, our chosen points are 'bad' if there exists some nonzero $w \in X$ such that

$$w(x_j) = 0 \qquad \text{for every} \quad j = 1, \ldots, k.$$

In other words, a collection of points is 'bad' if there is a nonzero element of X that is simultaneously zero at all these points. We use the analyticity of w to limit the size of its set of zeros (Theorem 15.4).

One can readily find (albeit artificial) examples of families of functions that are not real analytic for which a similar result fails. For example, if

$$u(x;\epsilon) = \begin{cases} 0 & -1 \leq x = 0, \\ \mathrm{e}^{-\epsilon/x^2} & 0 < x \leq 1 \end{cases}$$

and $\epsilon \in [1, 2]$ then no number of point observations within $[-1, 0]$ will serve to distinguish different members of this family.

15.1 Real analytic functions and the order of vanishing

15.1.1 Real analytic functions

We give here a very brief treatment of real analytic functions, following John (1982, Chapter 3.3), with proofs relegated to the exercises. Given a multi-index $\alpha = (\alpha_1, \ldots, \alpha_m)$, we write $\alpha! = \alpha_1! \cdots \alpha_m!$ and $x^\alpha = x_1^{\alpha_1} \cdots x_m^{\alpha_m}$. A function $f : \mathbb{R}^m \to \mathbb{R}$ is *real analytic at* x if there exists an $\epsilon > 0$ such that for some real coefficients $\{c_\alpha\}_{\alpha \geq 0}$ the equality

$$f(y) = \sum_{\alpha \geq 0} c_\alpha (y - x)^\alpha \tag{15.1}$$

holds for all y with $|y - x| \leq \epsilon$. A function $f : \mathbb{R}^m \to \mathbb{R}$ is *real analytic in* Ω, written $f \in C^\omega(\Omega)$, if f is real analytic at each $x \in \Omega$. A function

$f : \mathbb{R}^m \to \mathbb{R}^d$ is real analytic if each of its components is real analytic. The following theorem is a consequence of the results of Exercises 15.2 and 15.3.

Theorem 15.2 *If $f = (f_1, \ldots, f_d)$ is real analytic on an open set Ω then for any compact subset $K \subset \Omega$ there exist positive constants ϵ, M, and τ, such that for every $x \in K$,*

$$f(y) = \sum_{\alpha \geq 0} \frac{1}{\alpha!} D^\alpha f(x)(y-x)^\alpha \qquad \textit{for all } y \in \Omega \textit{ with } |y-x| < \epsilon, \quad (15.2)$$

and

$$|D^\beta f_k(x)| \leq M|\beta|!\tau^{-|\beta|} \qquad \textit{for all} \qquad k = 1, \ldots, d. \quad (15.3)$$

15.1.2 Order of vanishing

The order of vanishing of a C^∞ function $f : U \to \mathbb{R}$ at a point x is the smallest integer k such that $D^\alpha f(x) \neq 0$ for some multi-index α with $|\alpha| = k$. We say that f has finite order of vanishing in U if the order of vanishing of u is finite at every $x \in U$.

Lemma 15.3 *If U is a connected open subset of $\mathbb{R}^n$ and $f \in C^\omega(U)$ then f has finite order of vanishing in U.*

Proof We show that f is determined uniquely by its derivatives at any single point $x \in U$; then if f does not have finite order of vanishing in U, $D^\alpha f(x) = 0$ for every $\alpha \geq 0$ for some $x \in U$, from which it follows that $f \equiv 0$.

Fix some $x \in U$, and take $f, g \in C^\omega(U)$ with $D^\alpha f(x) = D^\alpha g(x)$ for every $\alpha \geq 0$. Let $h = f - g$, and define

$$U_1 = \{x \in U : D^\alpha h(x) = 0 \quad \text{for all } \alpha \geq 0\},$$
$$U_2 = \{x \in U : D^\alpha h(x) = 0 \quad \text{for some } \alpha \geq 0\}.$$

The set U_2 is open because $D^\alpha h$ is continuous for every $\alpha \geq 0$, and U_1 is also open, since if $D^\alpha h(x) = 0$ for every $\alpha \geq 0$, $h(y) = 0$ in a neighbourhood of x using (15.2). Since $x \in U_1$, it follows from the connectedness of U that U_2 is empty. □

In fact if f has finite order of vanishing in U, then its order of vanishing is uniformly bounded on any compact subset $K \subset U$. Arguing by contradiction, suppose not; then there is a sequence $x_j \in K$ with the order of vanishing of u at x_j at least j. Since K is compact, x_j has a subsequence that converges to some $x^* \in K$; it follows that u vanishes to infinite order at x^*, a contradiction. In particular, when we are dealing with real analytic periodic functions, the order of vanishing will be uniformly bounded, since we can restrict our attention to the fundamental compact domain $\Omega = \prod_{j=1}^m [0, L_j]$.

The following theorem, whose proof can be found in Kukavica & Robinson (2004), serves to limit the set of possible zeros of parametrised families of such functions.

Theorem 15.4 *Let K be a compact connected subset of $\mathbb{R}^m$. Suppose that for every fixed $p \in \Pi \subset \mathbb{R}^N$ the function $w = w(x; p)$,*

$$w: K \times \Pi \to \mathbb{R}^d,$$

has order of vanishing at most $j < \infty$, and is such that $\partial^\alpha w(x; p)$ depends on p in a θ-Hölder way for all $|\alpha| \leq j$. Then the zero set of $w(x; p)$, i.e.

$$\{(x, p): \ w(x, p) = 0\},$$

viewed as a subset of $K \times \Pi \subset \mathbb{R}^m \times \mathbb{R}^N$, is contained in a countable union of manifolds of the form

$$(x_i(x', p), x'; p),$$

where $x' = (x_1, \ldots, x_{i-1}, x_{i+1}, x_m)$ and x_i is a θ-Hölder function of its arguments.

15.2 Dimension and thickness of $\mathscr{A}$ in $C^r(\Omega, \mathbb{R}^d)$

We now show that $\mathscr{A}$ is a countable union of sets, all of which have box-counting dimension in $C^r(B, \mathbb{R}^d)$ (for any $r \in \mathbb{N}$) bounded by the box-counting dimension of $\mathscr{A}$ in $L^2(\Omega, \mathbb{R}^d)$ and all of which have thickness exponent zero in $C^r(B, \mathbb{R}^d)$ (for any $r \in \mathbb{N}$).

Since $\mathscr{A}$ consists of real analytic functions, and Ω is compact, for every element $u \in \mathscr{A}$ there exist $M > 0$ and $\tau > 0$ such that

$$|D^\alpha u(x)| \leq M|\alpha|!\tau^{-|\alpha|} \qquad \text{for all} \qquad x \in \Omega, \tag{15.4}$$

see (15.3). For $j \in \mathbb{N}$, we set

$$\mathscr{A}_j = \{u \in \mathscr{A} : \ u \text{ satisfies (15.4) with } M = j, \text{ and } \tau = 1/j\}; \tag{15.5}$$

clearly $\mathscr{A}_{j+1} \supseteq \mathscr{A}_j$, and $\mathscr{A} = \cup_{j=1}^\infty \mathscr{A}_j$.

Lemma 15.5 *For any $j \in \mathbb{N}$, for every $r \in \mathbb{N}$ the box-counting dimension of $\mathscr{A}_j$ in $C^r(\Omega, \mathbb{R}^d)$ is less than or equal to that of $\mathscr{A}$ in $L^2(\Omega, \mathbb{R}^d)$, and the thickness exponent of $\mathscr{A}_j$ in $C^r(\Omega, \mathbb{R}^d)$ is zero.*

Proof Standard Sobolev embedding results (see (10.1)) guarantee that

$$\|u\|_{C^r(\Omega,\mathbb{R}^d)} \le C\|u\|_{H^{r+(d/2)+1}(\Omega,\mathbb{R}^d)},$$

and if $A = -\Delta + I$, where Δ is the d-component Laplacian on Ω,

$$\|u\|_{H^{r+(d/2)+1}} \le c\|A^{(r+(d/2)+1)/2}u\|,$$

(see (10.3)), whence

$$\|u\|_{C^r(\Omega,\mathbb{R}^d)} \le c\|A^{(r+(d/2)+1)/2}u\|.$$

Thus, using the simple results of (3.5) and (7.2) (on box-counting dimension and thickness for sets considered as subsets of different spaces) it is sufficient to prove the lemma with $C^r(\Omega,\mathbb{R}^d)$ replaced by $D(A^r)$. We note here that since functions in $\mathscr{A}_j$ enjoy uniform bounds on their derivatives, $\mathscr{A}_j$ is uniformly bounded in $D(A^r)$ for each $r \in \mathbb{N}$,

$$\|A^r u\| \le R_r \qquad \text{for all} \qquad u \in \mathscr{A}_j. \tag{15.6}$$

We start with the box-counting dimension. If $s > r$, then for any $u \in D(A^s)$ we have the interpolation inequality

$$\|A^r u\| \le \|u\|^{1-(r/s)}\|A^s u\|^{r/s}$$

(see Exercise 13.4). It follows from (15.6) that the identity map from $\mathscr{A}_j$ onto itself is Hölder continuous as a map from $L^2(\Omega,\mathbb{R}^d)$ into $D(A^r)$ with Hölder exponent as close to 1 as we wish. That the box-counting dimension of $\mathscr{A}_j$ in the space $D(A^r)$ is bounded by $d_{\rm B}(\mathscr{A}_j;L^2) \le d_{\rm B}(\mathscr{A};L^2)$ is a consequence of part (iv) of Lemma 3.3.

In order to show that the thickness exponent is zero, let P_n denote the projection onto the space spanned by the first n eigenfunctions of A,

$$P_n u = \sum_{j=1}^{n}(u,w_j)w_j$$

and set $Q_n = I - P_n$. Recall that the nth eigenvalue of A satisfies $\lambda_n \sim n^{2/d}$ (see e.g. Davies (1995)). For any $k \in \mathbb{N}$ we have

$$\begin{aligned}|A^r Q_n u| = |A^{-k}Q_n(A^{k+r}u)| &\le \|A^{-k}Q_n\|_{\mathscr{L}(H)}|A^{k+r}u| \le \lambda_{n+1}^{-k}R_{k+r}\\ &\le C(n+1)^{-2k/d}R_{k+r} \le Cn^{-2k/d}R_{k+r}.\end{aligned}$$

Therefore,

$$\mathrm{dist}_{D(A^r)}\big(\mathscr{A}_j, P_n D(A^r)\big) \le \frac{CR_{k+r}}{n^{2k/d}},$$

and hence, using Exercise 7.1,

$$\tau(\mathscr{A}_j; D(A^r)) \leq \frac{d}{2k}.$$

Since this holds for any k, $\tau(\mathscr{A}_j; D(A^r)) = 0$ and the result follows. □

15.3 Proof of Theorem 15.1

We now give the proof of Theorem 15.1: recall that we have to show that almost every choice of k points $\mathbf{x} = (x_1, \ldots, x_k)$ from Ω makes the map

$$E_{\mathbf{x}}[u] = (u(x_1), \ldots, u(x_k))$$

one-to-one between $\mathscr{A}$ and its image.

Proof Suppose that almost every collection of k points makes $E_{\mathbf{x}}$ one-to-one on $\mathscr{A}_j$ (as defined in (15.5)) for each j. Then almost every collection of points is one-to-one on every $\mathscr{A}_j$, and hence on $\mathscr{A}$ itself: if

$$E_{\mathbf{x}}(u) = E_{\mathbf{x}}(v)$$

for some $u, v \in \mathscr{A}$ $(u \neq v)$, then $u, v \in \mathscr{A}_j$ for some j, and hence $u = v$. So we can fix j and concentrate on showing that almost every collection of k points makes $E_{\mathbf{x}}$ one-to-one on $\mathscr{A}_j$.

Let $W_j = (\mathscr{A}_j - \mathscr{A}_j)\backslash\{0\}$. If $E_{\mathbf{x}}$ is to be one-to-one on $\mathscr{A}_j$ then it should be nonzero on W_j. Since W_j consists of real analytic periodic functions, the order of vanishing of each $w \in W_j$ is uniformly bounded on Ω. Let

$$W_{j,r} = \{w \in W_j : \text{ the order of vanishing of } w \text{ is at most } r\}.$$

As above, if almost every $\mathbf{x}$ makes $E_{\mathbf{x}}$ one-to-one on $W_{j,r}$ for every $r \in \mathbb{N}$, then almost every $\mathbf{x}$ makes $E_{\mathbf{x}}$ one-to-one on W_j.

Lemma 15.5 implies that, for a fixed j and r,

$$d_{\mathrm{B}}(W_{j,r}; C^r(\Omega, \mathbb{R}^d)) \leq 2d_{\mathrm{B}}(\mathscr{A}; L^2) \qquad \text{and} \qquad \tau(W_{j,r}; C^r(\Omega, \mathbb{R}^d)) = 0.$$

Thus, using Proposition 7.10 (zero thickness implies zero dual thickness), $\tau^*(W_{j,r}; C^r(\Omega, \mathbb{R}^d)) = 0$, and hence Theorem 8.1 (embedding with Hölder continuous inverse) guarantees that for any

$$N > 4d_{\mathrm{B}}(\mathscr{A}) \qquad \text{and} \qquad \theta < 1 - (4d_{\mathrm{B}}(\mathscr{A})/N) \tag{15.7}$$

there is a parametrisation $w(x; p)$ of $W_{j,r}$ in terms of N coordinates $p \in \Pi \subset R^N$ which is θ-Hölder into $C^r(\Omega, \mathbb{R}^d)$. It follows that all the derivatives of w (with respect to x) up to order r depend in a θ-Hölder way on the parameter p.

Now, suppose that $\mathbf{x} = (x_1, \ldots, x_k)$ is a set of k points in Ω for which $E_\mathbf{x}$ is zero somewhere on $W_{j,r}$. Then there must exist a $p \in \Pi$ such that

$$w(x_i; p) = 0 \qquad \text{for all} \qquad i = 1, \ldots, k.$$

Theorem 15.4 guarantees that the zeros of w, considered as a subset of $\Omega \times \Pi$, are contained in a countable collection of sets, each of which is the graph of a θ-Hölder function,

$$(x', x_j(x'; \varepsilon); \varepsilon),$$

where $x' = (x_1, x_{j-1}, x_{j+1}, x_m)$. Each of these manifolds has $(m-1) + N$ free parameters.

It follows that collections of k such zeros (considered as a subset of $\Omega^k \times \Pi$) are contained in the product of k such manifolds. Since the coordinate p is common to each of these, they are the graphs of θ-Hölder functions from a subset of $\mathbb{R}^{N+(m-1)k}$ into $\mathbb{R}^k$. The result of Exercise 2.1 shows that each of these sets has Hausdorff dimension at most

$$N + (m-1)k + k(1-\theta),$$

and using the fact that the Hausdorff dimension is stable under countable unions (Proposition 2.8(iii)) the same goes for the whole countable collection.

The projection of this collection onto Ω^k enjoys the same bound on its dimension (since Lipschitz maps do not increase the Hausdorff dimension, Proposition 2.8(iv)), and so to make sure that these 'bad choices' do not cover $\Omega^k \subset \mathbb{R}^{mk}$ we need

$$N + (m-1)k + k(1-\theta) < mk.$$

This is certainly true if

$$k > \frac{N}{\theta}$$

and since the exponent θ can be chosen arbitrarily close to $1 - (4d_\mathrm{B}(\mathscr{A})/N)$ (see (15.7)), it follows that

$$k > \frac{N^2}{N - 4d_\mathrm{B}(\mathscr{A})}$$

will suffice. Choosing the integer value of N with $8d_\mathrm{B}(\mathscr{A}) - \frac{1}{2} \leq N < 8d_\mathrm{B}(\mathscr{A}) + \frac{1}{2}$ shows that $k \geq 16d_\mathrm{B}(\mathscr{A}) + 1$ is sufficient.

Since under this condition the collection of 'bad choices' is a subset of $\mathbb{R}^{km}$ with Hausdorff dimension less than km it follows from the fact that km-dimensional Hausdorff measure and Lebesgue measure on $\mathbb{R}^{km}$ are proportional (Theorem 2.4) that almost every choice of $\mathbf{x}$ (with respect to Lebesgue measure on Ω^k) makes $E_{\mathbf{x}}$ nonzero on $W_{j,r}$. □

15.4 Applications

15.4.1 Determining nodes

The theorem provides an instantaneous version of the 'determining nodes' introduced by Foias & Temam (1984): they called a collection of points $\{x_1, \ldots, x_k\}$ in Ω (asymptotically) 'determining' if for two solutions $u(x, t)$ and $v(x, t)$,

$$\max_{j=1,\ldots,k} |u(x_j, t) - v(x_j, t)| \to 0 \quad \text{as} \quad t \to \infty \tag{15.8}$$

implies that

$$\sup_{x \in \Omega} |u(x, t) - v(x, t)| \to 0 \quad \text{as} \quad t \to \infty.$$

Foias & Temam showed that for the two-dimensional Navier–Stokes equations there exists a δ such that if for every $x \in \Omega$

$$|x - x_j| < \delta \qquad \text{for some } j \in \{1, \ldots, k\}$$

then the collection of nodes is determining.[1] Under a mild additional condition our 'instantaneous determining nodes' are also asymptotically determining.

Lemma 15.6 *Suppose that the conditions of Theorem 15.1 hold, and that the attractor $\mathscr{A}$ attracts solutions in the norm of $L^\infty(\Omega)$. Then almost every set of k nodes $\{x_1, \ldots, x_k\}$ in Ω is asymptotically determining.*

Proof Since $\mathscr{A}$ is a compact subset of L^∞, the map $E_{\mathbf{x}}^{-1} : \mathbb{R}^{kd} \to L^\infty(\Omega)$ is continuous. Thus given any $\epsilon > 0$ there exists a δ, $0 < \delta < \epsilon$, such that for $u, v \in \mathscr{A}$,

$$\max_{j=1,\ldots,k} |u(x_j) - v(x_j)| < \delta \qquad \Rightarrow \qquad \|u - v\|_\infty < \frac{\epsilon}{3}.$$

Now let $u(t)$ and $v(t)$ be two solutions that agree asymptotically on the nodes $x_1, \ldots, x_k$ as in (15.8). Since $\mathscr{A}$ attracts in $L^\infty(\Omega)$, there exists a time $T > 0$

[1] In the same paper they conjectured that for solutions on the attractor coincidence of the values of u and v at k points (for k large enough) should imply coincidence of u and v; our theorem proves this conjecture when the attractor consists of real analytic functions. A proof for systems that possess an inertial manifold was given by Foias & Titi (1991).

such that for all $t \geq T$

$$\mathrm{dist}_{L^\infty}(u(x,t), \mathscr{A}) < \frac{\delta}{3}, \qquad \mathrm{dist}_{L^\infty}(v(x,t), \mathscr{A}) < \frac{\delta}{3}, \tag{15.9}$$

and also

$$\max_{j=1,\ldots,k} |u(x_j,t) - v(x_j,t)| < \frac{\delta}{3}.$$

It follows that there exist functions $u^*(t), v^*(t) \in \mathscr{A}$ such that

$$\|u^*(t) - u(t)\|_{L^\infty} < \frac{\delta}{3} \quad \text{and} \quad \|v^*(t) - v(t)\|_{L^\infty} < \frac{\delta}{3},$$

and consequently

$$\max_{j=1,\ldots,k} |u^*(x_j,t) - v^*(x_j,t)| < \delta.$$

It follows that $\|u^*(t) - v^*(t)\|_{L^\infty} < \epsilon/3$, which combined with (15.9) shows that $\|u(t) - v(t)\|_{L^\infty} \leq \epsilon$ for all $t \geq T$. □

15.4.2 Degrees of freedom in turbulent flows

Foias & Temam (1989) showed that if the forcing function f in the two-dimensional Navier–Stokes equations is real analytic then the attractor consists of real analytic functions. This means that as well as its interest as an abstract result, this theorem has application in the theory of turbulence (and more generally in any spatially extended system), allowing a rigorous connection between the attractor dimension and the 'number of degrees of freedom'.

Using dimensional analysis, Landau & Lifshitz (1959) introduced a heuristic notion of the 'number of degrees of freedom' in a turbulent fluid flow which has since been extensively applied. The result of their argument is that if l is 'the minimum significant length scale of the flow' (a quantity also arrived at via dimensional analysis), then the number of degrees of freedom of the flow is the number of boxes of side l needed to fill the domain Ω that contains the fluid, i.e. the 'number of degrees of freedom' in the flow should be defined as

$$\frac{\mathscr{L}^n(\Omega)}{l^n},$$

where $\Omega \subset \mathbb{R}^n$ – i.e. the number of 'little boxes of size l' that will fit into Ω.

If we identify 'the number of degrees of freedom of the flow' with the dimension of the global attractor (cf. Doering & Gibbon (1995)), then this suggests that the 'smallest significant length' will be given by

$$l \sim [d_{\mathrm{B}}(\mathscr{A})]^{-1/n}.$$

Theorem 15.1 allows the points $\{x_1, \ldots, x_k\}$ to be placed anywhere in the domain Ω, there just have to be a sufficient number $k \sim d_{\rm B}(A)$ of them. But if we decide to divide the space into equal boxes and place one node in each box, then the side of the box would have length $l \sim d_{\rm B}(A)^{-1/d}$. In this way Theorem 15.1 gives a rigorous derivation of Landau–Lifshitz heuristic.

It is particularly interesting to note that in the case of the two-dimensional Navier–Stokes equations with periodic boundary conditions, this gives an estimate for l that agrees with the results from the heuristic theory of two-dimensional turbulence due to Kraichnan (1976) which is the analogue of Kolmogorov's celebrated theory of three-dimensional turbulence. For more on the applications of these results to fluid dynamics see Robinson (2007).

Exercises

15.1 Let α, β be n-component multi-indices, and $x \in \mathbb{R}^n$ with $|x_i| < 1$ for all $i = 1, \ldots, n$. Show that

$$\sum_{\alpha:\ \alpha \geq \beta} \frac{\alpha!}{(\alpha - \beta)!} x^{\alpha - \beta} = \frac{\beta!}{(\underline{1} - x)^{\underline{1} + \beta}}, \tag{15.10}$$

where $\underline{1} = (1, \ldots, 1)$ (n times). [Hint: the left-hand side is $D^\beta \frac{1}{(\underline{1}-x)^{\underline{1}}}$.]

15.2 Suppose that $f : \mathbb{R}^n \to \mathbb{R}$ is real analytic at x with

$$f(y) = \sum_{\alpha \geq 0} c_\alpha (y - x)^\alpha \tag{15.11}$$

for all $|y - x| \leq \epsilon$. Fix $q \in (0, 1)$. Show that for any multi-index $\beta \geq 0$, for $|y - x| \leq q\epsilon$, the derivative $D^\beta f(y)$ can be obtained from term-by-term differentiation of (15.11), and

$$|D^\beta f(y)| \leq M |\beta|! \tau^{-|\beta|},$$

where

$$M = \frac{\mu}{(1 - q)^n} \qquad \text{and} \qquad \tau = (1 - q)\epsilon.$$

Deduce that $c_\alpha = (1/\alpha!) D^\alpha f(x)$.

15.3 Suppose that $f \in C^\omega(\Omega)$. Show that for any compact subset $K \subset \Omega$ there exist positive constants M and τ such that

$$|D^\beta f(x)| \leq M |\beta|! \tau^{-\beta}$$

for every $x \in K$.

Solutions to exercises

1.1 The closed set F_1 is contained in the open set $U_1 \cap (X \setminus \cap_{i=2}^{n+2} F_i)$, and so there exists an open set V_1 with

$$F_1 \subseteq V_1 \subseteq \overline{V_1} \subseteq U_1 \cap (X \setminus \cap_{i=2}^{n+2} F_i).$$

So $\overline{V_1} \subset U_1$ and $\overline{V_1} \cap \bigcap_{i=2}^{n+2} F_i = \emptyset$. Now, there exists an open set V_2 with

$$F_2 \subseteq V_2 \subseteq \overline{V_2} \subseteq U_2 \cap \left(X \setminus \left(\overline{V_1} \cap \bigcap_{i=3}^{n+2} F_i\right)\right),$$

and so $\overline{V_2} \subseteq U_2$ and $\overline{V_1} \cap \overline{V_2} \cap \bigcap_{i=3}^{n+2} F_i = \emptyset$. Continuing in this way shows that one can take the $\{F_i\}$ open in the original assumption.

Now let $\{U_1, \ldots, U_k\}$ be an open cover of X. If $k \leq n+1$ then this cover already has order $\leq n$, so we can assume that $k \geq n+2$. Set $V_j = U_j$ for $j = 1, \ldots, n+1$, and

$$V_{j+2} = \bigcup_{i=n+2}^{k} U_i.$$

These sets cover X, and so there exist open sets $\{F_i\}_{i=1}^{n+2}$ such that $F_i \subseteq V_i$, $X \subseteq \cup_{i=1}^{n+2} F_i$, and $\cap_{i=1}^{n+2} F_i = \emptyset$. Let $W_i = F_i$ for $i \leq n+1$ and $W_i = F_{n+2} \cap U_i$ for $i \geq n+2$. Then for every i, $W_i \subseteq U_i$,

$$X \subseteq \bigcup_{i=1}^{k} W_i, \qquad \text{and} \qquad \bigcap_{i=1}^{n+2} W_i = \emptyset.$$

One can perform the same construction for every subset of $\{1, \ldots, k\}$ consisting of $n+2$ elements, to deduce that every intersection of $n+2$ of the $\{W_i\}$ is empty, and hence that $\dim(X) \leq n$.

1.2 Let α_2 consist of all open unit squares in $\mathbb{R}^2$ of the form

$$(n, n+1) \times (m, m+1) \qquad n, m \in \mathbb{Z};$$

all the elements of A_2 are disjoint. Let α_1 consist of all the open edges of these squares,

$$\{n\} \times (m, m+1) \qquad \text{or} \qquad (n, n+1) \times \{m\},$$

but expanded to open subsets of $\mathbb{R}^2$, such that the resulting sets are pairwise disjoint. Let α_0 be the collection of all open balls of radius $\frac{1}{2}$ about the points (n, m) of the integer lattice. Then $\alpha = \alpha_0 \cup \alpha_1 \cup \alpha_2$ is a cover of $\mathbb{R}^2$ of mesh size 1 and of order 3, see the figure below.

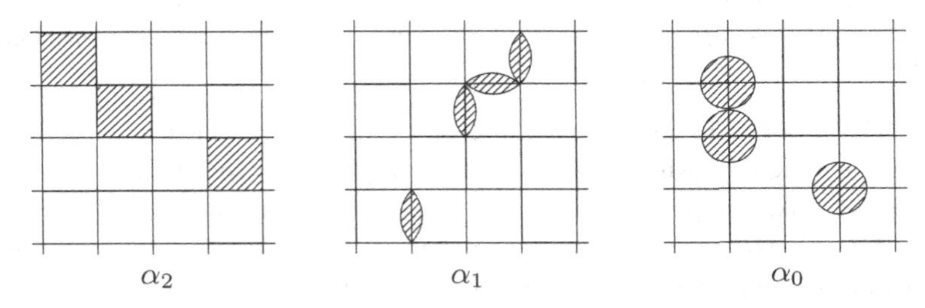

The three sets α_0, α_1, and α_2 that provide a cover of $\mathbb{R}^2$ of order 3.

If X is a compact subset of $\mathbb{R}^2$ then any covering β of X has Lebesgue number $\delta > 0$. Rescale the covering α by a factor of $\delta/3$, which produces a new covering of $\mathbb{R}^2$ of mesh size $\delta/3$ and order 3. The collection of all elements of this covering that intersect X forms a refinement of α of order 3, and so $\dim(X) \leq 2$.

1.3 Since K separates $A \cap C$ and $A \cap C'$ in A, there exist disjoint open sets U and U' such that

$$A \setminus K \subset U \cup U', \quad A \cap C \subset U, \quad \text{and} \quad A \cap C' \subset U'.$$

Note that since U and U' are open and disjoint, $\bar{U} \cap U' = \emptyset$.

Since C, C', and K are disjoint closed sets, there are open sets O_C, $O_{C'}$, and O_K that contain C, C', and K respectively and whose closures are disjoint. (Let X_1, X_2, and X_3 be closed subsets of X. Then there exist: (i) open sets U_1 and U_2 such that $X_1 \cup X_2 \subset U_1$, $X_3 \subset U_2$, and $\bar{U}_1 \cap \bar{U}_2 = \emptyset$; (ii) open sets V_1 and V_2 such that $X_1 \cup X_3 \subset V_1$, $X_2 \subset V_2$, and $\bar{V}_1 \cap \bar{V}_2 = \emptyset$; (iii) open sets W_1 and W_2 such that $X_1 \subset W_1$, $X_2 \cup X_3 \subset W_2$, and $\bar{W}_1 \cap \bar{W} = \emptyset$. We set $O_1 = U_1 \cap V_1 \cap W_1$, $O_2 = U_1 \cap V_2 \cap W_1$, and $O_3 = U_2 \cap V_1 \cap W_2$. Then each O_j

is open, $X_j \subset O_j$, and $\bar{O}_i \cap \bar{O}_j = \emptyset$ if $i \neq j$.) Set $W = O_C \cup U$; clearly $C \cup U \subset W$, and since $\bar{W} = \bar{O}_C \cup \bar{U}$, $\bar{W} \cap [C' \cup U'] = \emptyset$.

Now let $B = \partial W$. It follows that B separates C and C' in X; while if $x \in \bar{W} \setminus W$ then $x \in \bar{W}$, so $x \notin U'$, and $x \notin W$, so $x \notin U$. Thus if $x \in B \cap A$ it follows that $x \in K$.

1.4 Suppose that X is compact, $f : X \to Y$ is continuous and one-to-one, but $f^{-1} : f(X) \to X$ is not continuous. Then there exists an $\epsilon > 0$, a $y \in f(X)$, and a sequence $\{y_n\} \in f(X)$ such that $y_n \to y$ but

$$|f^{-1}(y_n) - f^{-1}(y)| > \epsilon. \tag{S.1}$$

However, $f^{-1}(y_n)$ is a sequence in the compact set, so it has a subsequence (which we relabel) such that $f^{-1}(y_n) \to x \in X$. Since f is continuous, it follows that $y_n \to f(x)$, so that $y = f(x)$. But then $x = f^{-1}(y)$, which contradicts (S.1).

1.5 The proof proceeds by induction on r; the result is clear if $r = 1$. So assume that the proposition holds for $r = k$; we wish to prove that it holds for $r = k + 1$. Take an open cover $\{U_1, \ldots, U_{k+1}\}$ of M, and let $U_k' = U_k \cup U_{k+1}$. Then $\{U_1, \ldots, U_{k-1}, U_k'\}$ is an open cover of M by k sets, and so by the induction hypothesis there exists an open covering $\{V_1, \ldots, V_{k-1}, V_{k'}\}$ of M such that

$$V_1 \subset U_1, \quad V_2 \subset U_2, \quad \cdots \quad V_{k-1} \subset U_{k-1}, \quad V_k' \subset U_k'$$

and the sets $\{V_1, \ldots, V_{k-1}, V_k'\}$ are mutually disjoint.

Now, $V_k' \cap M$ has dimension ≤ 0, and U_k and U_{k+1} cover $V_k' \cap M$. Thus there exist disjoint open sets V_k and V_{k+1} with $V_k \subset U_k$, $V_{k+1} \subset U_{k+1}$, and $V_k' \cap M \subseteq V_k \cup V_{k+1}$. It follows that $\{V_1, \ldots, V_{k+1}\}$ is a refinement of $\{U_1, \ldots, U_{k+1}\}$ consisting of disjoint open sets.

1.6 Write $X = \cup_{i=1}^{n+1} X_i$ where $\text{ind}(X_i) \leq 0$. If α is a cover of X then α is a cover of X_i for each i. Since $\text{ind}(X_i) \leq 0$ we can find a refinement of α, β_i, that covers X_i and consists of disjoint sets.

Now, $\beta = \cup_i \beta_i$ is a refinement of α that covers X. Any collection of more than $n + 2$ elements of β must contain two elements from one of the β_i, and hence there intersection is empty. So β is a refinement of order $\leq n + 1$.

1.7 Since the result of the previous exercise shows that $\dim(A) \leq \text{ind}(A)$, we only have to prove the reverse inequality. Take A with $\dim(A) \leq n$. Then by (i) A has a homeomorphic image that is a subset of $\mathscr{M}_{2n+1}^n \cap I_{2n+1}$. By (ii) $\text{ind}(\mathscr{M}_{2n+1}^n) = n$, and so by (iii) $\text{ind}(\mathscr{M}_{2n+1}^n \cap I_{2n+1}) \leq n$. Since $\text{ind}(\cdot)$ is a topological invariant, it follows that $\text{ind}(A) \leq n$.

2.1 Let $\epsilon > 0$. Since $d_{\mathrm{H}}(X) \leq n$, we can cover X by a collection $\{B(x_i, r_i)\}_{i \in I}$ of balls with centres $x_i \in X$ such that

$$\sum_{i \in I} r_i^{n+\epsilon} < \infty.$$

It follows that G is covered by the collection

$$\{B(x_i, r_i) \times B(f(x_i), Cr_i^{\theta})\}_{i \in I}.$$

Since $B(f(x_i), Cr_i^{\theta})$ is a subset of $\mathbb{R}^m$, we can cover it by m_i balls $B(y_{ij}, r_i)$, where $m_i \leq K(Cr_i^{\theta-1})^m + 1$ with K depending only on m. We therefore obtain

$$G \subseteq \bigcup_{i \in I} \bigcup_{j=1}^{m_i} B\big((f(x_i), y_{ij}), 2r_i\big).$$

Since

$$\begin{aligned}\sum_{i \in I} \sum_{j=1}^{m_i} (2r_i)^{n+(1-\theta)m+\epsilon} &= 2^{n+(1-\theta)m+\epsilon} \sum_{i \in I} m_i r_i^{n+(1-\theta)m+\epsilon} \\ &\leq K 2^{n+(1-\theta)m+\epsilon} C^m \sum_{i \in I} r_i^{n+\epsilon} + \sum_{i \in I} r_i^{n+(1-\theta)m+\epsilon} < \infty,\end{aligned}$$

we have $d_{\mathrm{H}}(G) \leq n + (1-\theta)m + \epsilon$, and since $\epsilon > 0$ is arbitrary the result follows.

2.2 Denote by $K_1, \ldots, K_m$ the closed intervals that make up

$$\Sigma_n := [0, \infty) \setminus (J_0 \cup \cdots \cup J_n).$$

Clearly $\Sigma_n \supset \Sigma$. The integers $\{n+1, n_2, \ldots\}$ can be partitioned into sets of indices $I_1, \ldots, I_m$ such that

$$K_j = \left(\bigcup_{q \in I_j} J_q \right) \cup (K_J \cap \Sigma).$$

Since the Lebesgue measure of Σ is zero, $K_j \cap \Sigma$ cannot contribute to the length of the interval K_j, and so, as the J_q, are disjoint, we must have

$$|K_j| = \sum_{q \in I_j} |J_q| \leq \epsilon, \tag{S.2}$$

using the first inequality in (2.6).

Now, Σ_n, and so Σ itself, can be covered by the intervals $K_1, \ldots, K_m$. Using (S.2) it follows that

$$\sum_{i=1}^{m} |K_i|^{1/2} = \sum_{i=1}^{m} \left(\sum_{q \in I_i} |J_q| \right)^{1/2} \leq \sum_{i=1}^{m} \sum_{q \in I_i} |J_q|^{1/2} = \sum_{q=n+1}^{\infty} |J_q|^{1/2} \leq \epsilon$$

using the second inequality in (2.6). Thus $\mathscr{H}^{1/2}(\Sigma) = 0$ as claimed.

2.3 Clearly $\mathscr{H}_\delta^d(X) \leq 2^d \mathscr{S}_\delta^d(X)$, since any cover by balls of radius δ provides a 2δ-cover of X. Also, if $\{U_i\}$ is a δ-cover of X then any U_i is contained in some ball of radius δ, so that $\mathscr{S}_\delta^d(X) \leq \mathscr{H}_\delta^d(X)$. It follows that $\mathscr{S}^d(X) \leq \mathscr{H}^d(X) \leq 2^d \mathscr{S}^d(X)$, and so the value of d at which $\mathscr{S}^d(X)$ jumps from ∞ to 0 is the same as that at which $\mathscr{H}^d(X)$ makes the same jump.

2.4 Let $M = \sup\{r(x) : \ x \in X\}$ and $A_1 = \{x \in A : \ 3M/4 < r(x) \leq M\}$. Choose some $x_1 \in A_1$, and then inductively

$$x_{k+1} \in A_1 \setminus \bigcup_{i=1}^{k} B(x_i, 3r(x_i)) \tag{S.3}$$

while the right-hand side of (S.3) is non-empty. The balls $\{B(x_i), r(x_i)\}$ are disjoint by definition, and lie in a compact subset of X, so there can be only a finite number of them, say k_1. Thus

$$A_1 \subseteq \bigcup_{i=1}^{k_1} B(x_i, r(x_i)).$$

Since $r(x) \leq 2r(x_i)$ for $x \in A_1$ and $i = 1 \ \ldots, k_1$, this implies that

$$\bigcup_{x \in A_1} B(x, r(x)) \subseteq \bigcup_{i=1}^{k_1} B(x_i, 5r(x_i)).$$

Now let

$$A_2 = \{x \in X : \ (\tfrac{3}{4})^2 M < r(x) \leq \tfrac{3}{4} M\}$$

and

$$A_2' = \{x \in A_2 : \ B(x, r(x)) \cap \bigcup_{i=1}^{k_1} B(x_i, r(x_i)) = \emptyset\}.$$

If $x \notin A_2 \setminus A_2'$ there is an $i \in \{1, \ldots, k_1\}$ such that $B(x, r(x)) \cap B(x_i, r(x_i)) \neq \emptyset$, and so

$$|x - x_i| \leq r(x) + r(x_i) \leq 3r(x_i).$$

Thus

$$A_2 \setminus A_2' \subseteq \bigcup_{i=1}^{k_1} B(x_i, 3r(x_i)). \tag{S.4}$$

Now pick $x_{k_1+1} \in A_2'$, and then choose inductively

$$x_{k+1} \in A_2' \setminus \bigcup_{i=k_1+1}^{k} B(x_i, 3r(x_i)).$$

As above, there exists a k_2 such that the balls $B(x_i, r(x_i))$, $i = 1, \ldots, k_2$ are disjoint and

$$A_2' \subseteq \bigcup_{i=k_1+1}^{k_2} B(x_i, 3r(x_i)).$$

Arguing as above, now using (S.4), we obtain

$$\bigcup_{x \in A_2} B(x, r(x)) \subseteq \bigcup_{i=1}^{k_2} B(x_i, 5r(x_i)).$$

Continuing this process gives the required disjoint subfamily.

2.5 Let V be any neighbourhood of S, and choose $\delta > 0$. For each $x \in S$, choose a ball $B_r(x)$ such that $B_r(x) \subset V$, $r < \delta$, and

$$\frac{1}{r^d} \iint_{B_r(x)} |f| > \delta.$$

Now, using the result of the previous exercise, find a disjoint subcollection of these balls $\{B_{r_i}(x_i)\}$ such that S is still covered by $\{B_{5r_i}(x_i)\}$. Since these balls are disjoint,

$$\iint_V |f| \geq \sum_i \iint_{B_{r_i}(x_i)} |f| \geq \delta \sum_i r_i^d.$$

Since $f \in L^1_{\text{loc}}(\Omega)$, the left-hand side is finite, so $\sum_i r_i^d \leq C$. Since S is contained in the union of $\{B_{5r_i}(x_i)\}$, and $r_i < \delta$ for every i, we must have

$$\mathscr{L}^n(S) \leq c \sum (5r_i)^n \leq c\delta^{n-d} \sum {r_i}^d \leq K\delta^{n-d}.$$

Since $\delta > 0$ was arbitrary, it follows that $\mathscr{L}^n(S) = 0$.

Since $|f|$ is integrable and V is an arbitrary neighbourhood of S (which has zero measure), we can make

$$\frac{1}{\delta} \iint_V |f|$$

as small as we wish by choosing V suitably. The above construction then furnishes a cover with $\sum_i r_i^d$ arbitrarily small, and so $\mathscr{H}^d(S) = 0$ as claimed.

3.1 In each case it suffices to show that there exist constants $c_1, c_2 > 0$ and $\alpha_1 \geq 1$, $\alpha_2 \leq 1$ such that

$$c_1 M(A, \alpha_1 \epsilon) \leq N(A, \epsilon) \leq c_2 M(A, \alpha_2 \epsilon), \tag{S.5}$$

since it follows from this that

$$\limsup_{\epsilon \to 0} \frac{\log M(A, \epsilon)}{-\log \epsilon} = \limsup_{\epsilon \to 0} \frac{\log N(A, \epsilon)}{-\log \epsilon}$$

(and similarly for the lim inf).

(i) It is clear that $N(A, \epsilon) \leq M(A, \epsilon)$. In order to prove the lower inequality in (S.5) consider a cover of A by $N(A, \epsilon)$ balls of radius ϵ, $B(x_i, \epsilon)$. Discarding any unnecessary balls from this cover, each ball $B(x_i, \epsilon)$ must contain a point $y_i \in A$. Since

$$B(y_i, 2\epsilon) \supset B(x_i, \epsilon)$$

it follows that $M(A, 2\epsilon) \leq N(A, \epsilon)$.

(ii) Let B_j, $j = 1, \ldots, M(A, \epsilon)$, be disjoint balls of radius ϵ with centres in A. Any $x \in A$ lies within ϵ of one of these balls, otherwise $B(x, \epsilon)$ would be an additional ball disjoint from the B_j, so $N(A, 2\epsilon) \leq M(A, \epsilon)$.

Conversely, given such a collection of disjoint balls and another collection B'_j of $\epsilon/2$-balls (not necessarily disjoint) that covers A, the centre of each B_j lies in one of the B'_j, and hence each B_j contains at least one of the $\epsilon/2$-balls, whence $M(A, \epsilon) \leq N(A, \epsilon/2)$.

(iii) In $\mathbb{R}^n$ any ϵ-ball is contained in at most 3^n boxes of side 2ϵ, while if X contains a point x in some ϵ-box, the whole box is contained in a ball of radius $\sqrt{n}\epsilon$ centred at x, from which it follows that

$$3^{-n} M(X, 2\epsilon) \leq N(X, \epsilon) \leq M(X, \epsilon/\sqrt{n}).$$

3.2 Take $d > d_{\rm B}(X)$. Then for ϵ sufficiently small X can be covered by ϵ^{-d} balls of radius ϵ centred in X. It follows that $O(X, \epsilon)$ can be covered by ϵ^{-d} balls of radius 2ϵ, and so $\mathscr{L}^n(O(X, \epsilon)) \leq \epsilon^{-d}(2\epsilon)^n \Omega_n$, from whence $c(X) \leq n - d$. For the opposite inequality, if $d < d_{\rm B}(X)$ then there is a sequence $\epsilon_j \to 0$ such that there are at least ϵ_j^{-d} disjoint balls of radius ϵ_j with centres in X: then $\mathscr{L}^n(O(X, \epsilon_j)) \geq \epsilon_j^{-d} \Omega_n \epsilon_j^n$, and $c(X) \geq n - d$.

3.3 Let $n = \dim(X)$, and let K_n consist of all mappings $f \in C(X, \mathbb{R}^{2n+1})$ such that $d_{\rm B}(f(X)) \leq n$. Now, by the characterisation of $d_{\rm LB}$ given in the hint,

$d_{\mathrm{LB}}(K) \leq n$ if and only if for every $k \in \mathbb{N}$, there exists a $\epsilon > 0$ such that $N_o(K, \epsilon) \leq \epsilon^{-n}/k$. Let $K_{n,k}$ be the class of all mappings f such that this inequality holds for some $\epsilon > 0$. Clearly $K_{n,k}$ is open, and

$$K_n = \bigcap_{k=1}^{\infty} K_{n,k}.$$

Now, as noted during the proof of the embedding theorem for sets with $\dim(X)$ finite, the embedding map g defined in (1.4) maps X into an n-dimensional polyhedron. Thus the set of maps $f \in C(X, \mathbb{R}^{2n+1})$ that map X into such a polyhedron is dense; for any such map,

$$d_{\mathrm{LB}}(f(X)) \leq d_{\mathrm{LB}}(\text{polyhedron}) \leq n.$$

K_n therefore contains a dense subset of $C(X, \mathbb{R}^{2n+1})$, so is itself dense.

We have shown that K_n is a dense G_δ in $C(X, \mathbb{R}^{2n+1})$, and Theorem 1.12 guarantees that the set of maps E_X that are embeddings of X is also a dense G_δ. It follows from the Baire Category Theorem that $K_n \cap E_X$ is also a dense G_δ, and in particular nonempty.

3.4 Find a set X' homeomorphic to X such that $\dim(X') = d_{\mathrm{B}}(X')$, and a set homeomorphic to Y such that $\dim(Y') = d_{\mathrm{B}}(Y')$. Then by Proposition 3.4, $d_{\mathrm{B}}(X' \times Y') \leq d_{\mathrm{B}}(X') + d_{\mathrm{B}}(Y')$. Since $X' \times Y'$ is homeomorphic to $X \times Y$ and dim is a topological invariant,

$$\begin{aligned} \dim(X \times Y) &= \dim(X' \times Y') \leq d_{\mathrm{B}}(X') + d_{\mathrm{B}}(Y') \\ &= \dim(X') + \dim(Y') = \dim(X) + \dim(Y). \end{aligned}$$

3.5 Choose $s > d_{\mathrm{H}}(X)$ and $t > d_{\mathrm{B}}(Y)$. Then there exists a $\delta_0 > 0$ such that $N(Y, \delta) \leq \delta^{-t}$ for all $\delta \leq \delta_0$. Let $\{B(x_i, r_i)\}$ be a cover of X such that

$$\sum_i r_i^s < 1,$$

which is possible since $\mathscr{H}^s(X) = 0$ for $s > d_{\mathrm{H}}(X)$. Now for each i cover Y with $N_i := N(Y, r_i)$ balls of radius r_i, $\{B(y_{i,j}, r_i)\}_{j=1}^{N_i}$. Thus

$$X \times Y \subset \bigcup_i \bigcup_j B(x_i, r_i) \times B(y_j, r_i).$$

Thus

$$\begin{aligned}\mathcal{H}^{s+t}_{2\delta}(X \times Y) &\leq \sum_i \sum_j (2r_i)^{s+t} \\ &\leq \sum_i N(y, r_i) 2^{s+t} r_i^{s+t} \\ &\leq 2^{s+t} \sum_i r_i^{-t} r_i^{s+t} < 2^{s+t}.\end{aligned}$$

It follows that $\mathcal{H}^{s+t}(X \times Y) < \infty$, and hence that $d_H(X \times Y) \leq s + t$. Since $s > d_H(X)$ and $t > d_B(Y)$ were arbitrary, this completes the proof.

3.6 Let $X \subseteq \cup_{i=1}^{\infty} X_i$ with each X_i closed. Then using the Baire Category Theorem there is an index j and an open set $U \subset \mathbb{R}^n$ such that $X \cap U \subset X_j$. So $d_B(X) = d_B(X_j)$. It follows using the definition that $d_{MB}(X) \geq d_B(X)$, and we already have the reverse inequality.

3.7 If $X \subseteq \cup_i X_i$ then

$$d_P(X) \leq \sup_i d_P(X_i) \leq \sup_i d_B(X_i).$$

It follows from the definition of d_{MB} that $d_P(X) \leq d_{MB}(X)$. Conversely, suppose that $s > \dim_P(X)$. Then $\mathcal{P}^s(X) = 0$, and so we can find a collection of sets X_i such that $X \subset \cup_i X_i$ with $\mathcal{P}^s_0(X_i) < \infty$ for each i. In particular, $N_\delta(X_i)\delta^s$ is bounded as $\delta \to 0$ for each i, from which it follows that $d_B(X_i) \leq s$ for each i, and hence $d_{MB}(X) \leq s$.

5.1 Take a probe space E of constant functions

$$E = \{g_c \in L^1(0, 1) : \; g_c(x) = c \quad \text{for all} \quad x \in [0, 1], \; 0 \leq c \leq 1\},$$

i.e. a set isometric to [0, 1], equipped with Lebesgue measure. Then

$$\int f(x) + g_c(x)\,\mathrm{d}x = \int f(x)\,\mathrm{d}x + c,$$

which is zero for at most one $c \in [0, 1]$. Thus $\int f + g \neq 0$ for almost every $g_c \in E$.

5.2 To show that E is compact it suffices to consider only one 'component' of L, i.e. to prove the compactness of E_0. Given a sequence $l^{(n)} \in E_0$ with

$$l^{(n)} = \sum_{j=1}^{\infty} j^{-\gamma} [\phi_j^{(n)}]^*, \quad \phi_j^{(n)} \in S_j,$$

since each S_j is compact one can extract successive subsequences and then use a diagonal argument to find a subsequence (which we relabel) such that for every j, $\phi_j^{(n)} \to \phi_j$ as $n \to \infty$. It is then straightforward to show that

$$l^{(n)} \to l = \sum_{j=1}^{\infty} j^{-\gamma} \phi_j^*,$$

where clearly $l \in E_0$.

5.3 Note that for each s one can view K_s as a subset of $\mathbb{R}^{d_j-1}$, and that the set $(1-t)K_a + tK_b$ is precisely the intersection of the convex hull of K_a and K_b with $\Pi + ((1-t)a + tb)\gamma$, and so in particular is a subset of $K_{(1-t)a+b}$. It follows from the Brunn–Minkowski inequality that

$$\mathscr{L}^{d_j-1}(K_{(1-t)a+b})^{1/(d_j-1)} \geq (1-t)\mathscr{L}^{d_j-1}(K_a)^{1/(d_j-1)} + t\mathscr{L}^{d_j-1}(K_b)^{1/(d_j-1)},$$

i.e. that the map $s \mapsto \mathscr{L}^{d_j-1}(K_s)^{1/(d_j-1)}$ is concave. Since U_j is symmetric this map is also symmetric, and hence it attains its maximum value when $s = 0$.

7.1 Denote the right-hand side of (7.4) by $\tilde{\tau}$. Taking any $\sigma \in (0, \tau(X))$, there is a sequence $\epsilon_j \in (0,1)$ converging to 0 such that

$$d(X, \epsilon_j) > \epsilon_j^{-\sigma} \geq \left\lfloor \epsilon_j^{-\sigma} \right\rfloor = n_j,$$

where $\lfloor x \rfloor$ denotes the integer part of x. Since $\varepsilon(X, n_j) \geq \epsilon_j$,

$$\frac{\log n_j}{-\log \varepsilon(X, n_j)} \geq \frac{\log \lfloor \epsilon_j^{-\sigma} \rfloor}{\log(1/\epsilon_j)} \geq \frac{\log(\epsilon_j^{-\sigma} - 1)}{\log(1/\epsilon_j)},$$

which shows that $\tilde{\tau} \geq \sigma$. Since $\sigma < \tau$ was arbitrary, one can conclude that $\tau(X) \leq \tau$.

7.2 For any $d > d_{\mathrm{B}}(X)$, there exists an ϵ_0 such that for any $\epsilon < \epsilon_0$ one can cover X by no more than $N_\epsilon = \epsilon^{-d}$ balls of radius ϵ, with centres $\{x_j\}_{j=1}^{N_\epsilon}$. Use the Johnson–Lindenstrauss Lemma to find a function $f : H \to \mathbb{R}^n$, where $n = O(\ln N_\epsilon)$, such that

$$\frac{1}{2}\|x_i - x_j\| \leq |f(x_i) - f(x_j)| \leq 2\|x_i - x_j\| \qquad \text{for all} \qquad i, j = 1, \ldots, N_\epsilon.$$

The mapping $f^{-1}|_{\{f(x_1),\ldots,f(x_N)\}}$ is 2-Lipschitz onto $\{x_1, \ldots, x_N\}$. In particular, it can be extended to a 2-Lipschitz map from $\mathbb{R}^n$ into H. Since any $x \in X$ lies within ϵ of one of the $\{x_j\}$, this shows that there exists a 2-Lipschitz mapping $\varphi : \mathbb{R}^n \to H$ such that

$$\mathrm{dist}(X, \varphi(\mathbb{R}^n)) \leq \epsilon.$$

Since $n = O(\ln N_\epsilon) = O(-d \ln \epsilon)$, it follows that $\tau_{\mathrm{LE}}(X) = 0$.

7.3 Let $\mu = \det(e_1, \ldots, e_n)$, and define $f_j : \mathbb{R}^n \to \mathbb{R}$ for $1 \le j \le n$ by

$$f_j(x) = \frac{1}{\mu} \det(e_1, \ldots, e_{j-1}, x, e_{j+1}, \ldots, e_n).$$

Clearly $f_j(e_k) = \delta_{jk}$, f_j is linear, and hence, by the choice of $(e_1, \ldots, e_n)$, $\|f_j\|_* \le 1$. Since $f_j(e_j) = 1$, it follows that $\|f_j\|_* = 1$.

7.4 Let $\{e_1, \ldots, e_n\}$ be an Auerbach basis for U, and $\{f_1, \ldots, f_n\}$ the associated elements of U^* such that $\|f_i\|_{U^*} = 1$ and $f_i(e_j) = \delta_{ij}$. Extend each f_i to an element $\phi_i \in \mathscr{B}^*$ with $\|\phi_i\|_{\mathscr{B}^*} = 1$, and define

$$Pu = \sum_{j=1}^{n} \phi_j(u) e_j.$$

Then P is a projection onto U with

$$\|Pu\| \le \sum_{j=1}^{n} \|\phi_j(u) e_j\| \le \sum_{j=1}^{n} |\phi_j(u)| \|e_j\| \le \sum_{j=1}^{n} \|u\| = n\|u\|,$$

i.e. $\|P\| \le n$. (A significantly more involved argument due to Kadec & Snobar (1971) provides a projection whose norm is no larger than $(\dim(U))^{1/2}$, see also Proposition 12.14 in Meise & Vogt, 1997).

8.1 Given any rank k orthogonal projection P_0, choose an orthonormal basis $\{e_1, \ldots, e_k\}$ for $P_0 H$ and by identifying $x \in P_0 H$ with the coefficients of x in its expansion in terms of this basis define a linear map $M_0 : P_0 H \to \mathbb{R}^k$; note that for $u \in P_0 H$, $|M_0 u| = \|u\|$. Set $L_0 = M_0 P_0$ so that $L_0 \in \mathscr{L}(H, \mathbb{R}^k)$. Since prevalence implies density (see comment after Lemma 5.2), given any $\epsilon > 0$ there exists a linear map $L \in E$ with $\|L\| < \epsilon$ such that $L_0 + L$ is injective on X and satisfies

$$\|x - y\| \le C\|(L_0 + L)(x - y)\|^\theta \qquad \text{for all} \qquad x, y \in X.$$

Using Lemma 6.1, $L_0 + L = MP$, where P is an orthogonal projection of rank k. If $Pu = 0$ then

$$\|P_0 u\| = \|M_0 P_0 u\| = \|L_0 u\| = \|MPu - Lu\| = \|Lu\| \le \epsilon\|u\|,$$

and so it follows that $\|P - P_0\| < \epsilon$.

8.2 For each $n \in \mathbb{N}$ the set

$$\mathbb{X}_n = \bigcup_{|j| \le n} X_j$$

has $d_{\rm B}(\mathbb{X}_n) \le d$, and hence $\tau^*(\mathbb{X}_n) \le d$ (Lemma 7.9). It follows from Theorem 8.1 that if $k > 2d$ and

$$0 < \theta < \frac{k - 2d}{k(1+d)}$$

there is a prevalent set of maps $L : \mathscr{B} \to \mathbb{R}^k$ that are injective on $\mathbb{X}_n$ and satisfy

$$\|x - y\| \le C_{L,n}|Lx - Ly|^\theta \qquad \text{for all} \qquad x, y \in \mathbb{X}_n. \tag{S.6}$$

Since the countable intersection of prevalent sets is prevalent, there is a prevalent set of mappings $L : \mathscr{B} \to \mathbb{R}^k$ that satisfy (S.6) for every $n \in \mathbb{N}$. In particular, these mappings are injective on $\bigcup_j X_j$, since if $x, y \in \bigcup_j X_j$, then $x, y \in \mathbb{X}_n$ for some n, and L is injective on $\mathbb{X}_n$.

8.3 Let $\alpha_j = a_j e_j$, where the $\{e_j\}$ are orthonormal. If L is bi-Lipschitz on X there exists a $C > 0$ such that

$$\|L(a_j e_j) - L(0)\| = |a_j|\|Le_j\| \ge C\|\alpha_j\| = C|a_j|,$$

i.e. $\|Le_j\| \ge C$. Using Lemma 6.1 one can write any $L : H \to \mathbb{R}^k$ as $L = MP$, with P an orthogonal projection of rank k. It follows that $\|Pe_j\| \ge C'$, and so using Lemma 6.3,

$$\text{rank}(P) \ge \sum_{j=1}^{\infty} C' = \infty.$$

9.1 Expanding the norms as inner products yields

$$\begin{aligned} \|x\|^2 - (x, y) - (x, z) + \frac{1}{4}\|y\|^2 + \frac{1}{4}\|z\|^2 + \frac{1}{2}(y, z) \\ \le \frac{1}{2}\|x\|^2 + \frac{1}{2}\|y\|^2 - (x, y) + \frac{1}{2}\|x\|^2 \\ + \frac{1}{2}\|z\|^2 - (x, z) + \frac{1}{4}(\|y\|^2 + \|z\|^2 - 2(y, z)); \end{aligned}$$

on cancelling terms

$$0 \le \frac{1}{2}(\|y\|^2 + \|z\|^2 - 2(y, z)).$$

Applying (9.13) to $\{x_0, x_1, x_3\}$ and $\{x_2, x_1, x_3\}$ we obtain

$$\begin{aligned} \left\| x_0 - \frac{x_1 + x_3}{2} \right\|^2 &\le \frac{1}{2}\|x_0 - x_1\|^2 + \frac{1}{2}\|x_0 - x_3\|^2 - \frac{1}{4}\|x_1 - x_3\|^2, \\ \left\| x_2 - \frac{x_1 + x_3}{2} \right\|^2 &\le \frac{1}{2}\|x_2 - x_1\|^2 + \frac{1}{2}\|x_2 - x_3\|^2 - \frac{1}{4}\|x_1 - x_3\|^2, \end{aligned}$$

and adding these gives

$$\left\|x_0 - \frac{x_3 + x_1}{2}\right\|^2 + \left\|x_2 - \frac{x_3 + x_1}{2}\right\|^2 + \frac{1}{2}\|x_1 - x_3\|^2 \leq \frac{1}{2}\sum_{j=0}^{3} \|x_j - x_{j+1}\|^2;$$

and (9.14) follows using the triangle inequality. (In fact this inequality holds in any CAT(0) space, see Sato (2009).)

9.2 First note that

$$|s_j(x)| = |\varrho(x, x_j) - \varrho(x_j, x_0)| \leq \varrho(x, x_0),$$

and so $s(x) \in \ell^\infty$. Then

$$|s_j(x) - s_j(y)| = |\varrho(x, x_j) - \varrho(y, x_j)|,$$

from which it follows immediately that $\|s(x) - s(y)\|_{\ell^\infty} \leq \varrho(x, y)$. The lower bound $\|s(x) - s(y)\|_{\ell^\infty} \geq \varrho(x, y)$ follows using the fact that $\{x_j\}$ is dense: in particular for any $\epsilon > 0$ there exists a $j \in \mathbb{N}$ such that $\varrho(y, x_j) < \epsilon$, and hence

$$|s_j(x) - s_j(y)| \geq \varrho(x, y) - 2\varrho(y, x_j) > \varrho(x, y) - 2\epsilon.$$

10.1 Since $f(x) = \sum_k c_k \mathrm{e}^{\mathrm{i}kx}$ it follows that

$$\begin{aligned}
|f(x)| &\leq \sum_k |c_k| \leq \sum \frac{1}{1 + |k|^2}^{1/2} (1 + |k|^2)^{1/2} |c_k| \\
&\leq \left(\sum_k \frac{1}{1 + |k|^2}\right)^{1/2} \left(\sum_k (1 + |k|^2)|c_k|^2\right)^{1/2} \\
&\leq C\|u\|_{H^1}^2.
\end{aligned}$$

Since $\sum_{|k| \leq n} c_k \mathrm{e}^{\mathrm{i}kx}$ is continuous for each n, f is the uniform limit of continuous functions, so continuous.

10.2 Take a sequence $\{u_n\}_{n=1}^\infty$ with u_n bounded in $D(A^\beta)$. Let

$$u_n = \sum_{j=1}^{\infty} c_{n,j} w_j, \qquad \text{so that} \qquad \|A^\beta u_n\|^2 = \sum_{j=1}^{n} \lambda_j^{2\beta} |c_{n,j}|^2 \leq M$$

for some $M > 0$. It follows that for each j, $\{c_{n,j}\}_{n=1}^\infty$ is a bounded sequence of real numbers, one can find a succession of subsequences and then use the standard diagonal method to find a subsequence (which we relabel) such that for every n,

$$c_{n,j} \to c_n^* \qquad \text{as} \qquad j \to \infty.$$

Let $u = \sum_{n=1}^{\infty} c_n^* w_n$; note that $\sum_n \lambda_n^{2\beta} |c_n^*|^2 \leq M$. Now,

$$\begin{aligned}
\|A^\alpha(u_n - u)\|^2 &= \sum_{n=1}^{\infty} \lambda_j^{2\alpha} |c_{n,j} - c_n^*|^2 \\
&\leq \sum_{n=1}^{m} \lambda_j^{2\alpha} |c_{n,j} - c_n^*|^2 + \lambda_{m+1}^{-2(\beta-\alpha)} \sum_{n=m+1}^{\infty} \lambda_j^{2\beta} |c_{n,j} - c_n^*|^2 \\
&\leq \sum_{n=1}^{m} \lambda_j^{2\alpha} |c_{n,j} - c_n^*|^2 + 2M\lambda_{m+1}^{-2(\beta-\alpha)}.
\end{aligned}$$

Given $\epsilon > 0$, choose m sufficiently large that the second term is $\leq \epsilon/2$; one can then choose j large enough to ensure that the first term (with involves only a finite number of coefficients) is also $\leq \epsilon/2$.

10.3 Since

$$\frac{\mathrm{d}}{\mathrm{d}\lambda}(\lambda^\gamma \mathrm{e}^{-\lambda t}) = \gamma \lambda^{\gamma-1} \mathrm{e}^{-\lambda t} - t\lambda^\gamma \mathrm{e}^{-\lambda t} = \lambda^\gamma \mathrm{e}^{-\lambda t}[\gamma\lambda^{-1} - t],$$

$\lambda^\gamma \mathrm{e}^{-\lambda t}$ attains its maximum when $\lambda = \gamma/t$. Thus

$$\max_{\lambda \geq \lambda_1} \lambda^\gamma \mathrm{e}^{-\lambda t} = \begin{cases} \gamma^\gamma \mathrm{e}^{-\gamma} t^{-\gamma} & 0 < t < \gamma/\lambda_1 \\ \lambda_1^\gamma \mathrm{e}^{-\lambda_1 t} & t \geq \gamma/\lambda_1, \end{cases}$$

from which (10.7) follows. Now

$$\begin{aligned}
\int_0^\infty \|A^\gamma \mathrm{e}^{-At}\|_{\mathscr{L}(H)} \,\mathrm{d}t &\leq \int_0^{\gamma/\lambda_1} \gamma^\gamma \mathrm{e}^{-\gamma} t^{-\gamma} \,\mathrm{d}t + \int_{\gamma/\lambda_1}^\infty \lambda_1^\gamma \mathrm{e}^{-\lambda_1 t} \,\mathrm{d}t \\
&\leq \gamma^\gamma \mathrm{e}^{-\gamma} \frac{1}{1-\gamma} \left(\frac{\gamma}{\lambda_1}\right)^{1-\gamma} + \lambda_1^{\gamma-1} \mathrm{e}^{-\gamma} \\
&= \lambda_1^{\gamma-1} \frac{\mathrm{e}^{-\gamma}}{1-\gamma}.
\end{aligned}$$

10.4 From the variation of constants formula

$$u(t) = \mathrm{e}^{-At} u_0 + \int_0^t \mathrm{e}^{-A(t-s)} g(u(s)) \,ds,$$

and so

$$A^\beta u(t) = A^\beta \mathrm{e}^{-At} u_0 + \int_0^t A^\beta \mathrm{e}^{-A(t-s)} g(u(s)) \,\mathrm{d}s.$$

Using (10.7) and (10.8)

$$\|A^{\beta}u(t)\| \leq \|A^{\beta-\alpha}\mathrm{e}^{-At}\|_{\mathscr{L}(H)}\|A^{\alpha}u_0\| + \int_0^t \|A^{\beta}\mathrm{e}^{-A(t-s)}\|_{\mathscr{L}(H)}\|A^{\alpha}u(s)\|\,\mathrm{d}s$$
$$\leq t^{\alpha-\beta}\|A^{\alpha}u_0\| + I_{\beta}\sup_{0\leq s\leq t}\|A^{\alpha}u(s)\| < \infty.$$

10.5 Taking the inner product of $\mathrm{d}u/\mathrm{d}t + Au + B(u,u) = 0$ with u yields

$$\frac{1}{2}\frac{\mathrm{d}}{\mathrm{d}t}\|u\|^2 + \|Du\|^2 = 0,$$

using $(Au, u) = \|Du\|^2$ (10.16) and $(B(u,u), u) = 0$ (10.17). Now use the Poincaré inequality (10.19) $\|u\| \leq \|Du\|$, so that

$$\frac{1}{2}\frac{\mathrm{d}}{\mathrm{d}t}\|u\|^2 = -\|Du\|^2 \leq -\|u\|^2.$$

It follows that

$$\|u(t)\|^2 \leq \|u_0\|^2\mathrm{e}^{-2t},$$

and so $\|u(t)\| \to 0$ as $t \to \infty$.

10.6 Given $u = \sum_{k\in\dot{\mathbb{Z}}^2} c_k\mathrm{e}^{\mathrm{i}k\cdot x}$,

$$D_j u = \sum_{k\in\dot{\mathbb{Z}}^2} \mathrm{i}k_j c_k \mathrm{e}^{\mathrm{i}k\cdot x} \qquad \text{implies that} \qquad \|D_j u\|^2 = \sum_{k\in\dot{\mathbb{Z}}^2}|k_j|^2|c_k|^2.$$

Thus

$$\|Du\|^2 = \sum_{j=1}^{2}\|D_j u\|^2 = \sum_{k\in\dot{\mathbb{Z}}^2}|k|^2|c_k|^2 \geq \sum_{k\in\dot{\mathbb{Z}}^2}|c_k|^2 = \|u\|^2.$$

11.1 Let X be compact and invariant. Since X is compact it is bounded, so it is attracted to $\mathscr{A}$. Therefore

$$\mathrm{dist}(S(t)X, \mathscr{A}) = \mathrm{dist}(X, \mathscr{A}) \to 0 \qquad \text{as} \qquad t \to \infty,$$

i.e. $\mathrm{dist}(X, \mathscr{A}) = 0$ so $X \subseteq \mathscr{A}$. Similarly, if Y attracts all bounded sets then Y attracts $\mathscr{A}$, and so

$$\mathrm{dist}(S(t)\mathscr{A}, Y) = \mathrm{dist}(\mathscr{A}, Y) \to 0 \qquad \text{as} \qquad t \to \infty,$$

i.e. $\mathrm{dist}(\mathscr{A}, Y) = 0$ so that $\mathscr{A} \subseteq Y$.

11.2 Clearly if $x = \lim_{n\to\infty} S(t_n)b_n$, $t_n \to \infty$, and $b_n \in B$ then

$$x \in \overline{\bigcup_{s\geq t} S(s)B}$$

for all $t \geq 0$, since the right-hand side is closed. Conversely, if

$$x \in \bigcap_{t\geq 0} \overline{\bigcup_{s\geq t} S(s)B}$$

then for any sequence $t_n \to \infty$

$$x \in \overline{\bigcup_{s\geq t_n} S(s)B}.$$

For any n such that $x \in \bigcup_{s\geq t_n} S(s)B$, there exists a τ_n and b_n such that $x = S(\tau_n)b_n$. Otherwise

$$x = \lim_{j\to\infty} S(s_j)b_j$$

for some sequences s_j and $b_j \in B$. If s_j is unbounded then we are done; if s_j is bounded then one can find a τ_n and $b_n \in B$ such that $\|x - S(\tau_n)b_n\| < 1/n$.

11.3 Suppose that $\mathscr{A}$ is not connected. Then there exist open sets O_1 and O_2 such that $O_1 \cap \mathscr{A} \neq \emptyset$ and $O_2 \cap \mathscr{A} \neq \emptyset$,

$$O_1 \cup O_2 \supset \mathscr{A}, \qquad \text{and} \qquad O_1 \cap O_2 = \emptyset.$$

Since the compact attracting set $K \subset B(0, R)$ for some $R > 0$, and a ball in $\mathscr{B}$ is connected, it follows that $S(t)B(0, R)$, the continuous image of $B(0, R)$, is connected. Since $\mathscr{A}$ attracts $B(0, R)$, for t sufficiently large, $S(t)B(0, R)$ is contained either wholly in O_1 or wholly in O_2. Since $B(0, R) \supset K$, $\omega(B(0, R)) \supset \omega(K)$ and hence $\mathscr{A} = \omega(B(0, R))$ is contained in either O_1 or O_2, contradicting our initial assumption.

11.4 For any $u_0 \in U(X)$, $u_0 = S(t)u(-t)$; since $u(t) \to X$ as $t \to -\infty$, it follows that $\gamma_- = \cup_{t\leq 0} u(t)$ is bounded. Thus

$$\operatorname{dist}(u_0, \mathscr{A}) = \operatorname{dist}(S(t)u(-t), \mathscr{A}) \leq \operatorname{dist}(S(t)\gamma_-, X) \to 0$$

as $t \to \infty$, and hence $u_0 \in \mathscr{A}$.

11.5 Rearrange the governing inequality and multiply by the integrating factor $\exp(-\int_s^t a(u)\,\mathrm{d}u)$:

$$\frac{\mathrm{d}}{\mathrm{d}t}\left(\mathrm{e}^{-\int_s^t a(u)\,\mathrm{d}u} x(t)\right) \leq b(t)\mathrm{e}^{-\int_s^t a(u)\,\mathrm{d}u} \leq b(t).$$

Integrate this inequality from s to $t + r$, so that

$$\mathrm{e}^{-\int_s^{t+r} a(u)\,\mathrm{d}u} x(t+r) \leq x(s) + \int_s^{t+r} b(u)\,\mathrm{d}u \leq x(s) + B,$$

i.e.

$$x(t+r) \le e^{\int_s^{t+r} a(u)\,du}[x(s)+B] \le e^A[x(s)+B].$$

Now integrate again from t to $t+r$ with respect to s to obtain

$$rx(t+r) \le e^A[X+rB],$$

and hence $x(t+r) \le e^A[B+(X/r)]$.

11.6 (i) Returning to (11.9), we retain the term in $\|Au\|^2$ to give

$$\frac{d}{dt}\|Du\|^2 + \|Au\|^2 \le \|f\|^2.$$

Integrating from 0 to 1 then yields

$$\|Du(1)\|^2 + \int_0^1 \|Au(s)\|^2\,ds \le \|f\|^2 + \|Du_0\|^2 \le 7\|f\|^2.$$

(ii) Using the triangle inequality on (11.6),

$$\begin{aligned}\|u_t\| &\le \|Au\| + \|B(u,u)\| + \|f\| \\ &\le \|Au\| + c_1\|u\|^{1/2}\|Du\|\|Au\|^{1/2} + \|f\| \\ &\le \frac{3}{2}\|Au\| + \frac{c_1^2}{2}\|u\|\|Du\|^2 + \|f\|,\end{aligned}$$

whence

$$\begin{aligned}\|u_t\|^2 &\le \frac{9}{2}\|Au\|^2 + \frac{c_1^4}{2}\|u\|^2\|Du\|^4 + 2\|f\|^2 \\ &\le \frac{9}{2}\|Au\|^2 + 36c_1^4\|f\|^6 + 2\|f\|^2.\end{aligned}$$

It follows that

$$\int_0^1 \|u_t(s)\|^2\,ds \le \frac{9}{2}(7\|f\|^2) + \left(36c_1^4\|f\|^6 + 2\|f\|^2\right) =: I_t(c_1, \|f\|).$$

(iii) Differentiating (11.6) with respect to t we obtain

$$\frac{du_t}{dt} + Au_t + B(u,u_t) + B(u_t,u) = 0.$$

Taking the inner product of this with u_t yields

$$\begin{aligned}\frac{1}{2}\frac{d}{dt}\|u_t\|^2 + \|Du_t\|^2 &= -(B(u_t,u),u_t) \\ &\le c_2\|u_t\|\|Du_t\|\|Du\| \\ &\le \frac{1}{2}\|Du_t\|^2 + \frac{c_2^2}{2}\|u_t\|^2\|Du\|^2,\end{aligned}$$

using the orthogonality property (10.17) and the inequality (11.11). Since $\|Du(t)\|^2 \leq 6\|f\|^2$ it follows that

$$\frac{\mathrm{d}}{\mathrm{d}t}\|u_t\|^2 + \|Du_t\|^2 \leq c_2^2\|u_t\|^2\|Du\|^2 \leq 6c_2^2\|f\|^2\|u_t\|^2.$$

Dropping the $\|Du_t\|^2$ and integrating with respect to t from s to 1 yields

$$\|u_t(1)\|^2 \leq \|u_t(s)\|^2 + 6c_2^2\|f\|^2 \int_s^1 \|u_t(r)\|^2\,\mathrm{d}r,$$

and integrating once more from 0 to 1 with respect to s we obtain

$$\|u_t(1)\|^2 \leq (1 + 6c_2^2\|f\|^2) \int_0^1 \|u_t(s)\|^2\,\mathrm{d}s \leq (1 + 6c_2^2\|f\|^2)I_t =: \rho_t^2.$$

(iv) Once again we use the triangle inequality on (11.6), this time to obtain

$$\begin{aligned}\|Au\| &\leq \|u_t\| + \|B(u,u)\| + \|f\| \\ &\leq \|u_t\| + c_1\|u\|^{1/2}\|Du\|\|Au\|^{1/2} + \|f\| \\ &\leq \|u_t\| + \frac{c_1^2}{2}\|u\|\|Du\|^2 + \frac{1}{2}\|Au\| + 2\|f\|,\end{aligned}$$

whence

$$\begin{aligned}\|Au(1)\| &\leq 2\|u_t(1)\| + c_1^2\|u(1)\|\|Du(1)\|^2 + 2\|f\| \\ &\leq \left[2\rho_t + 6\sqrt{2}c_1^2\|f\|^3 + \|f\|\right] =: \rho_A.\end{aligned}$$

Since $\mathscr{A}$ is invariant, if $u_1 \in \mathscr{A}$ there exists $u_0 \in \mathscr{A}$ such that $u_1 = S(1)u_0$, and hence $\|Au_1\| = \|AS(1)u_0\| \leq \rho_A$ for every $u_1 \in \mathscr{A}$.

12.1 Clearly, for each $\lambda > 0$ and $x \in K$, $\mathrm{D}f(x) \in \mathscr{L}_{\lambda/2}(X)$ for all $\lambda > 0$ and $\nu_\lambda(\mathrm{D}f(x)) = \nu(x)$. Consequently, for each $0 < \lambda < \frac{1}{2}$,

$$d_{\mathrm{B}}(K) \leq \nu\,\frac{\log\left((\nu+1)\frac{D}{\lambda}\right)}{\log(1/2\lambda)}.$$

Taking the limit as $\lambda \to 0$ we obtain $d_{\mathrm{B}}(K) \leq \nu$.

12.2 For $Y(t)$ to be a supersolution of $\dot{y} = a + b\int_0^t (t-s)^\alpha y(s)\,\mathrm{d}s$ we require

$$2a\mathrm{e}^{Kt} \geq a + 2ab\int_0^t (t-s)^{-\alpha}\mathrm{e}^{Ks}\,\mathrm{d}s,$$

i.e.

$$2 \geq \mathrm{e}^{-Kt} + 2b\int_0^t (t-s)^{-\alpha}\mathrm{e}^{-K(t-s)}\,\mathrm{d}s = \mathrm{e}^{-Kt} + 2bK^{-(1-\alpha)}\int_0^t u^{-\alpha}\mathrm{e}^{-u}\,\mathrm{d}u.$$

Since $\Gamma(z) = \int_0^\infty t^{z-1}\mathrm{e}^{-t}\,\mathrm{d}t$, this is certainly ensured if

$$2 \geq \mathrm{e}^{-Kt} + \frac{2b\Gamma(1-\alpha)}{K^{1-\alpha}}.$$

So it suffices to choose $K = (2b\Gamma(1-\alpha))^{1/(1-\alpha)}$. (In the case considered in this exercise, this argument, due to Robinson (1997), offers a significantly simpler proof than that due to Henry (1981, Lemma 7.1.1); while Henry's result is sharper, the bound here is often sufficient in applications.)

12.3 Using the bounds on $\|A^\alpha \mathrm{e}^{-At}\|$ in (10.9) and (10.10), there exists a constant $c > 0$ such that

$$\|A^\alpha \mathrm{e}^{-At} Q_n\| \leq ct^{-\alpha}\mathrm{e}^{-(\lambda_{n+1}-1)t} \qquad \text{for all} \qquad t \geq 0.$$

Therefore

$$\begin{aligned}
&\|Q_n \mathrm{D}S(t;u_0)\|_{\mathscr{L}(D(A^\alpha))} \\
&\qquad \leq \mathrm{e}^{-\lambda_{n+1}t} + cM\int_0^t (t-s)^{-\alpha}\mathrm{e}^{-(\lambda_{n+1}-1)(t-s)}\|\mathrm{D}S(s;u_0)\|_{\mathscr{L}(D(A^\alpha))}\,\mathrm{d}s.
\end{aligned}$$

Using (12.10), for all $0 \leq t \leq 1$,

$$\begin{aligned}
\|Q_n \mathrm{D}S(t;u_0)\|_{\mathscr{L}(D(A^\alpha))} &\leq \mathrm{e}^{-\lambda_{n+1}t} + cKM\int_0^t (t-s)^{-\alpha}\mathrm{e}^{-(\lambda_{n+1}-1)(t-s)}\,\mathrm{d}s \\
&\leq \mathrm{e}^{-\lambda_{n+1}t} + cKM\int_0^t u^{-\alpha}\mathrm{e}^{-(\lambda_{n+1}-1)u}\,\mathrm{d}u,
\end{aligned}$$

and since $\Gamma(z) = \int_0^\infty t^{z-1}\mathrm{e}^{-t}\,\mathrm{d}t$,

$$\|Q_n \mathrm{D}S(1;u_0)\|_{\mathscr{L}(D(A^\alpha))} \leq \mathrm{e}^{-\lambda_{n+1}} + \frac{cKM\Gamma(1-\alpha)}{(\lambda_{n+1}-1)^{1-\alpha}}. \tag{S.7}$$

12.4 If $\bar{\omega}_d^{1/d} < \alpha_1(\mathrm{D}f(u))$ then, using Lemmas 12.8 and 12.9, the number of balls of radius $\sqrt{2}\bar{\omega}_d^{1/d}$ needed to cover $\mathrm{D}f(u)[B(0,1)]$ is bounded by

$$\frac{4^j\omega_j(\mathrm{D}f(u))}{\bar{\omega}_d^{j/d}},$$

where j is the largest integer such that $\bar{\omega}_d^{j/d} \leq \alpha_j$. Since $\bar{\omega}_d^{1/d} \geq \bar{\alpha}_d$, it follows that $j \leq d$. So no more than

$$\max_{1\leq j\leq d} \frac{4^j\omega_j(\mathrm{D}f(u_j))}{\bar{\omega}_d^{j/d}} \leq \max_{1\leq j\leq d} 4^j\frac{\bar{\omega}_j}{\bar{\omega}_d^{j/d}} =: M$$

balls of radius $\sqrt{2}\bar{\omega}_d^{1/d}$ are required to cover $\mathrm{D}f(u)[B(0,1)]$.

Alternatively, if $\bar{\omega}_d^{1/d} \geq \alpha_1(\mathrm{D}f(u))$ then

$$\mathrm{D}f(u)[B(0,1)] \subseteq B(0, \alpha_1(\mathrm{D}f(u))) \subseteq B(0, \bar{\omega}_d^{1/d}),$$

and it requires only one ball of radius $\sqrt{2}\bar{\omega}_d^{1/d}$ to cover $\mathrm{D}f(u)[B(0,1)]$ in this case.

Thus $\mathrm{D}f(u)[B(0,1)]$ can always be covered by M balls of radius $2\bar{\omega}_d^{1/d}$.

12.5 We need to have $\theta < 1$ in order to apply Lemma 12.1. Using the hint repeatedly, note that for each $u \in K$

$$\omega_j(\mathrm{D}[f^k](u)) \leq \omega_j(\mathrm{D}f(f^{k-1}(u))) \cdots \omega_j(\mathrm{D}f(u)) \leq \bar{\omega}_j^k;$$

if we consider f^k rather than f we can replace $\bar{\omega}_j$ by $\bar{\Omega}_j := \bar{\omega}_j^k$. Thus given d and γ such that (12.12) holds, we can find a k sufficiently large such that

$$2\bar{\Omega}_d^{1/d} = 2\bar{\omega}_d^{k/d} < 1$$

and

$$\begin{aligned}(2\bar{\Omega}_d^{1/d})^\gamma \max_{1\leq j\leq d} 4^j \frac{\bar{\Omega}_j}{\bar{\Omega}_d^{j/d}} &= (2\bar{\omega}_d^{k/d})^\gamma \max_{1\leq j\leq d} 4^j \frac{\bar{\omega}_j^k}{\bar{\omega}_d^{kj/d}} \\ &\leq 2^\gamma 4^d \left[\bar{\omega}_d^\gamma \max_{1\leq j\leq d} \frac{\bar{\omega}_j^d}{\bar{\omega}_d^j}\right]^{k/d} < 1. \qquad \text{(S.8)}\end{aligned}$$

We now apply Lemma 12.1, making use of the observation in the footnote.

12.6 Since q_j is concave, there exist α, β such that $q_j \leq -\alpha j + \beta$: choose α and β such that $0 < q_{n-1} = -\alpha(n-1) + \beta$ and $0 > q_n = -\alpha n + \beta$. In particular it follows that $\beta/\alpha < n$. The argument above leading to the lower bound on γ uses only upper bounds on the $\bar{q}_j$s, so $d_{\mathrm{B}}(\mathscr{A}) \leq \gamma$ provided that

$$\gamma > \max_{1\leq j\leq d} \frac{j(-\alpha d + \beta) - d(-\alpha j + \beta)}{-\alpha d + \beta} = \max_{1\leq j\leq d} \frac{\beta(d-j)}{\alpha d - \beta} \leq \frac{\beta d}{\alpha d - \beta}.$$

Since d is arbitrary, one can let $d \to \infty$ and show that $d_{\mathrm{B}}(\mathscr{A}) \leq \gamma$ provided that $\gamma > \beta/\alpha$. But $\beta/\alpha < n$, so $d_{\mathrm{B}}(\mathscr{A}) \leq n$.

12.7 Denote by w_j the eigenfunction corresponding to λ_j, and expand each ϕ_j in terms of the eigenbasis $\{w_j\}$ to obtain

$$\sum_{j=1}^n \|A^{1/2}\phi_j\|^2 = \sum_{j=1}^n \sum_{k=1}^\infty \lambda_k |(\phi_j, w_k)|^2 = \sum_{k=1}^\infty \lambda_k \left(\sum_{j=1}^n |(\phi_j, w_k)|^2\right).$$

Since $\|\phi_j\| = 1$ we have $\sum_{j=1}^n \sum_{k=1}^\infty |(w_k, \phi_j)|^2 = n$, and since the $\{\phi_j\}$ are orthonormal we have $\sum_{j=1}^n |(w_k, \phi_j)|^2 \leq 1$, from which the result follows.

12.8 Given a set $\{\phi_j\}_{j=1}^n$ that is orthonormal in H, for any $u \in \mathscr{A}$ we have

$$\begin{aligned}
\sum_{j=1}^n (\phi_j, -A\phi_j + \mathrm{D}g(u)\phi_j) &\le -\sum_{j=1}^n \|A^{1/2}\phi_j\|^2 + M \sum_{j=1}^n \|A^\alpha \phi_j\| \|\phi_j\| \\
&\le -\sum_{j=1}^n \|A^{1/2}\phi_j\|^2 + M \sum_{j=1}^n \|A^{1/2}\phi_j\|^{2\alpha} \|\phi_j\|^{2-2\alpha} \\
&\le -\sum_{j=1}^n \|A^{1/2}\phi_j\|_{1/2}^2 + M \left(\sum_{j=1}^n \|A^{1/2}\phi_j\|^2 \right)^\alpha n^{1-\alpha} \\
&\le (1-\alpha) \left[-\sum_{j=1}^n \|A^{1/2}\phi_j\|^2 + M^{1/(1-\alpha)} n \right] \\
&\le (1-\alpha) \left[-\sum_{j=1}^n \lambda_j + M^{1/(1-\alpha)} n \right].
\end{aligned}$$

It follows from (12.13) that the final line provides an upper bound for $q_n(\mathrm{D}S(1; u_0))$ which is uniform over all $u_0 \in \mathscr{A}$. Since this bound is a concave function of n, it follows from Exercise 12.6 that $d_{\mathrm{B}}(\mathscr{A}) \le n$ once the right-hand side is negative, which gives (12.15).

13.1 Write $w = u - v$. Then

$$\|A^\beta w\|^2 = \|A^\beta (P_n w + Q_n w)\|^2 = \|A^\beta P_n w\|^2 + \|A^\beta Q_n w\|^2 \ge \lambda_{n+1}^{2\beta} \|Q_n w\|^2$$

and

$$L^2 \|w\|^2 = L^2 \|P_n w\|^2 + L^2 \|Q_n w\|^2.$$

Whence

$$(\lambda_{n+1}^{2\beta} - L^2) \|Q_n w\|^2 \le L^2 \|P_n w\|^2.$$

If n is large enough that $\lambda_{n+1}^{2\beta} > 2L^2$ then

$$\|Q_n(u - v)\| \le \|P_n(u - v)\| \qquad \text{for all} \qquad u, v \in \mathscr{A}. \tag{S.9}$$

Now define $\phi : P_n\mathscr{A} \to Q_n H$ by $\phi(P_n u) = Q_n u$ for all $u \in \mathscr{A}$. The inequality (S.9) shows that this is well defined and 1-Lipschitz where defined; as in the proof of Theorem 13.3 this function can be extended to a 1-Lipschitz function $\Phi : P_n H \to Q_n H$.

13.2 If $y(0) \leq (\delta/\gamma)^{1/2}$ then clearly $y(t) \leq (\delta/\gamma)^{1/2}$ for all $t \geq 0$. If $y(0) \geq (\delta/\gamma)^{1/2}$ then there exists $t_0 \in (0, \infty)$ such that

$$y(t) \geq (\delta/\gamma)^{1/2} \qquad \text{for} \qquad 0 \leq t \leq t_0$$

and

$$y(t) \leq (\delta/\gamma)^{1/2} \qquad \text{for} \qquad t \geq t_0.$$

For $t \in [0, t_0]$ consider $z(t) = y(t) - (\delta/\gamma)^{1/2} \geq 0$; then

$$y^2 = (z + (\delta/\gamma)^{1/2})^2 \geq z^2 + (\delta/\gamma),$$

and so

$$\dot{z} + \gamma z^2 \leq \dot{y} + \gamma \left(y^2 - \frac{\delta}{\gamma} \right) \leq 0.$$

Integrating $\dot{z} + \gamma z^2 \leq 0$ yields

$$z(t) \leq \frac{1}{z_0^{-1} + \gamma t} \leq \frac{1}{\gamma t}.$$

This implies (13.15) for $t \in [0, t_0]$, and since $y(t) \leq (\delta/\gamma)^{1/2}$ for all $t \geq t_0$ the result follows.

13.3 The eigenvalues are the sums of two square integers (positive and negative); so we will have reached $2k^2$ once we have taken $[2(k-1)]^2 + 1$ combinations of integers with modulus $\leq k$. So if $4(k-1)^2 < n \leq 4k^2$ then

$$2k^2 \leq \lambda_n < 2(k+1)^2;$$

since $k - 1 \leq (n/4)^{1/2} \leq k$,

$$\tfrac{1}{2}n^{1/2} \leq k < k + 1 < n^{1/2},$$

and so $\frac{1}{2}n \leq \lambda_n < 2n$.

13.4 Put $u = \sum_{j=1}^{\infty} c_j w_j$. Then

$$\begin{aligned}\|A^\alpha u\|^2 &= \sum_{j=1}^{\infty} \lambda_j^{2\alpha} |(u, w_j)|^2 \\ &= \sum_{j=1}^{\infty} \lambda_j^{2\alpha} |c_j|^{2\alpha/\beta} |c_j|^{2(1-(\alpha/\beta))} \\ &\le \left(\sum_{j=1}^{\infty} \lambda_j^{2\beta} |c_j|^2 \right)^{\beta/\alpha} \left(\sum_{j=1}^{\infty} |c_j|^2 \right)^{1-(\alpha/\beta)} \\ &= \|A^\beta u\|^{\alpha/\beta} \|u\|^{1-(\alpha/\beta)},\end{aligned}$$

using Hölder's inequality with exponents $(\alpha/\beta, 1/(1-\alpha/\beta))$.

13.5 Fix an $n \ge n_0$, and let X be a subset of $\mathscr{A}$ that is maximal for the relation

$$\|Q_n(u-v)\|_\alpha \le \|P_n(u-v)\|_\alpha \qquad \text{for all} \qquad u, v \in X.$$

As in the proof of Theorem 13.3(ii), it follows that there exists a 1-Lipschitz function $\Phi_n : P_n H \to Q_n H$ such that $X \subset G_{P_n H}[\Phi_n]$.

If $u \in \mathscr{A}$ but $u \notin X$ then there is a $v \in X$ such that

$$\|Q_n(u-v)\|_\alpha \ge \|P_n(u-v)\|_\alpha. \tag{S.10}$$

Since $S(t^*)\mathscr{A} = \mathscr{A}$ (because $\mathscr{A}$ is invariant) there exist $\bar{u}, \bar{v} \in \mathscr{A}$ such that $u = S(t^*)\bar{u}$ and $v = S(t^*)\bar{v}$; since (S.10) implies that (13.16) cannot hold, it follows from (13.17) that

$$\|u - v\| \le \delta_n \|\bar{u} - \bar{v}\| \le 2M\delta_n.$$

Since $v \in G_{P_n H}[\Phi_n]$ this implies that $\mathscr{A}$ lies within a $4M\delta_n$ neighbourhood of $G_{P_n H}[\Phi_n]$ as claimed.

13.6 Given a $u^* \in H^2 \setminus H^3$, set $f = Au^* + B(u^*, u^*)$. Then $f \in L^2$ and u^* is a stationary solution of the equations. The attractor must contain u^*, and hence cannot be bounded in H^3.

13.7 Assume that $\|w(0)\| \neq 0$; we will show that $\|w(t)\| \neq 0$ for any $t \ge 0$. Taking the inner product of (13.7) with w yields

$$\frac{1}{2}\frac{\mathrm{d}}{\mathrm{d}t}\|w\|^2 + \|A^{1/2} w\|^2 = (w, h(t)).$$

Dividing by $L(t)\|w\|^2$ we obtain

$$\frac{1}{2L(t)\|w\|^2}\frac{\mathrm{d}}{\mathrm{d}t}\|w\|^2 + \tilde{Q}(t) = \frac{(w, h(t))}{L(t)\|w\|^2} \ge -\frac{1}{L(t)} - \frac{\|h\|^2}{L(t)\|w\|^2},$$

and using (13.9) this gives

$$\frac{1}{2L(t)\|w\|^2}\frac{\mathrm{d}}{\mathrm{d}t}\|w\|^2 + \tilde{Q}(t) \geq -1 - 2K\tilde{Q}^{2\alpha} \geq -1 - 2\alpha\tilde{Q} - (1-2\alpha),$$

using Young's inequality ($ab \leq (a^p/p) + (b^q/q)$ when $p^{-1} + q^{-1} = 1$). Thus

$$\frac{1}{2L(t)\|w\|^2}\frac{\mathrm{d}}{\mathrm{d}t}\|w\|^2 + c\tilde{Q}(t) \geq -c'.$$

Since $\tilde{Q}$ is bounded, this inequality is simply

$$\frac{1}{2}\frac{\mathrm{d}}{\mathrm{d}t}(-\log L(t)) \geq -C,$$

and so

$$-\log L(t) + \log L(0) \geq -2Ct.$$

Thus $\log L(t) \leq 2Kt + \log L(0)$, and hence $\|w(t)\| \neq 0$.

14.1 If $M^T Me = \lambda e$ then

$$MM^T(Me) = M[M^T Me] = M[\lambda e] = \lambda[Me]$$

and if $MM^T\hat{e} = \lambda\hat{e}$ then

$$M^T[MM^T\hat{e}] = M^T M[M^T\hat{e}] = M^T[\lambda\hat{e}] = \lambda[M^T\hat{e}].$$

14.2 For any $z \in \mathbb{R}^N$, write

$$Lz = \sum_{\alpha,j} c_{\alpha,j}(z, e_\alpha)\hat{e}_j = M_z\underline{c},$$

where $\underline{c} \in \mathbb{R}^{Nk}$ with components $c_{\alpha,j}$, and M_z is a transformation from $\mathbb{R}^{Nk}$ into $\mathbb{R}^k$ with components

$$[M_z]_{i,\{\alpha,j\}} = (z, e_\alpha)\delta_{ij} \qquad \alpha = 1, \ldots, N;\ i, j = 1, \ldots, k.$$

In order to apply Lemma 14.3 we need to find the singular values of M_z. We calculate these by considering $M_z M_z^T$ rather than $M_z^T M_z$, since

$$\begin{aligned}[M_z M_z^T]_{r,s} &= \sum_{\alpha,j}[M_z]_{r,\{\alpha,j\}}[M_z]_{s,\{\alpha,j\}} \\ &= (z, e_\alpha)\delta_{sj}(z, e_\alpha)\delta_{rj} = |z|^2\delta_{rs} :\end{aligned}$$

MM^T has k nonzero singular values, all of which are $|z|$.

14.3 Given $x \in \mathbb{R}^N$ and $L \in E$, write $Lx = M_x \underline{c}$ with M_x as in Exercise 14.2 and $\underline{c} \in \mathbb{R}^{Nk}$ depending on L. Lemma 14.3 shows that

$$\frac{\mathrm{Vol}\{\underline{c} \in B_{Nk}) : |M_x \underline{c}| < \delta\}}{\mathrm{Vol}(B_{Nk})} \le C_{Nk,k} \left(\frac{\delta}{|x|} \right)^k,$$

since the kth largest singular value of M_x is k, using the result of the previous exercise.

14.4 We have

$$\begin{aligned}
\Phi(\mu) &\sim \frac{\mu^\alpha}{\mu - \cdots} \left[\frac{2\mu - 2\mu^2 + \frac{4}{3}\mu^3 - \cdots}{2\mu} - \frac{(\mu - \frac{1}{2}\mu^2 + \frac{1}{6}\mu^3 - \cdots)^2}{\mu^2} \right]^{1/2} \\
&= \frac{\mu^\alpha}{\mu - \cdots} \left[(1 - \mu + \tfrac{2}{3}\mu^2 - \cdots) - (1 - \tfrac{1}{2}\mu + \tfrac{1}{6}\mu^2)^2 \right]^{1/2} \\
&= \frac{\mu^\alpha}{\mu - \cdots} \left[(1 - \mu + \tfrac{2}{3}\mu^2 - \cdots) - (1 - \mu + \tfrac{1}{4}\mu^2 + \tfrac{1}{3}\mu^2 + \ldots) \right]^{1/2} \\
&= \frac{\mu^\alpha}{\mu - \cdots} [\tfrac{1}{12}\mu^2 + \cdots]^{1/2} \sim \frac{\mu^\alpha}{2\sqrt{3}}.
\end{aligned}$$

15.1 Using the series expansion of $(1 - \xi)^{-1}$ for $\xi \in \mathbb{R}$ repeatedly,

$$\frac{1}{(\underline{1} - x)^{\underline{1}}} = \prod_{i=1}^{n} \sum_{j \ge 0} x_n^j = \sum_{\alpha \ge 0} x^\alpha.$$

Applying D^β to both sides yields (15.10).

15.2 Since $\sum_{\alpha \ge 0} c_\alpha (y - x)^\alpha$ converges for $|y - x| \le \epsilon$, it follows that

$$\mu := \sum_{\alpha \ge 0} |c_\alpha| \epsilon^{|\alpha|} < \infty,$$

and in particular $|c_\alpha| \epsilon^{|\alpha|} \le \mu$ for every $\alpha \ge 0$.

Now take q with $0 < q < 1$. Then for any y with $|y - x| \le q\epsilon$,

$$\begin{aligned}
\sum_{\alpha \ge 0} |D^\beta c_\alpha x^\alpha| &\le \sum_{\alpha \ge \beta} \frac{\alpha!}{(\alpha - \beta)!} |c_\alpha| \, q^{|\alpha - \beta|} \epsilon^{|\alpha - \beta|} \\
&\le \frac{\mu}{\epsilon^{|\beta|}} \sum_{\alpha \ge \beta} \frac{\alpha!}{(\alpha - \beta)!} \underline{q}^{\alpha - \beta} \\
&= \frac{\mu}{\epsilon^{|\beta|}} \frac{\beta!}{(1 - q)^{n + |\beta|}},
\end{aligned}$$

using (15.10). It follows that for $|y - x| \leq q\epsilon$, $D^\beta f(y)$ is continuous and given by

$$D^\beta f(y) = \sum_{\alpha \geq \beta} \frac{\alpha!}{(\alpha - \beta)!} c_\alpha (y - x)^{\alpha - \beta}.$$

In particular, for any multi-index β,

$$|D^\beta f(y)| \leq M|\beta|!\tau^{-|\beta|},$$

where

$$M = \frac{\mu}{(1 - q)^n} \qquad \text{and} \qquad \tau = (1 - q)\epsilon.$$

Finally, differentiating (15.1) at $y = x$ shows that $c_\alpha = (1/\alpha!)D^\alpha f(x)$.

15.3 For every $x \in \Omega$ there are positive numbers $M(x)$, $\tau(x)$, and $\epsilon(x) > 0$ such that

$$|D^\beta f(y)| \leq M(x)|\beta|!\tau(x)^{-|\beta|} \qquad \text{for all} \qquad |y - x| < \epsilon(x).$$

If K is a compact subset of Ω, a finite number of the balls $\{B(x, \epsilon(x))\}$ covers K, say $\{B(x_j, \epsilon(x_j))\}_{j=1}^N$. Then for every $x \in K$,

$$|D^\beta f(x)| \leq M|\beta|!\tau^{-|\beta|},$$

where $M = \max_j M(x_j)$ and $\tau = \min_j \tau(x_j)$.

References

Assouad, A. (1983) Plongements lipschitziens dans $\mathbb{R}^n$. *Bull. Soc. Math. France* **111**, 429–448.

Ball, K. (1986) Cube slicing in $\mathbb{R}^n$. *Proc. Amer. Math. Soc.* **97**, 465–473.

Ben-Artzi, A., Eden, A., Foias, C., & Nicolaenko, B. (1993) Hölder continuity for the inverse of Mañé's projection. *J. Math. Anal. Appl.* **178**, 22–29.

Benyamini, Y., & Lindenstrauss, J. (2000) *Geometric Nonlinear Functional Analysis*, Vol. 1., AMS Colloquium Publications 48 (Providence, RI: American Mathematical Society).

Boichenko, V. A., Leonov, G. A., & Reitmann, V. (2005) *Dimension Theory for Ordinary Differential Equations* (Wiesbaden: Teubner).

Bollobás, B. (1990) *Linear Analysis* (Cambridge: Cambridge University Press).

Bouligand, M. G. (1928) Ensembles impropres et nombre dimensionnel. *Bull. Sci. Math.* **52**, 320–344 and 361–376.

Caffarelli, L., Kohn, R., & Nirenberg, L. (1982) Partial regularity of suitable weak solutions of the Navier–Stokes equations. *Comm. Pure Appl. Math.* **35**, 771–831.

Carvalho, A. N., Langa, J. A., & Robinson, J. C. (2010) Finite-dimensional global attractors in Banach spaces. *J. Diff. Eq.* **249**, 3099–3109.

Charalambous, M. G. (1999) A note on the relations between fractal and topological dimensions. *Q & A in General Topology*. **17**, 9–16.

Chepyzhov, V. V. & Ilyin, A. A. (2004) On the fractal dimension of invariant sets: applications to Navier–Stokes equations. *Discr. Cont. Dynam. Syst.* **10**, 117–135.

Chepyzhov, V. V., & Vishik, M. I. (2002) *Attractors for Equations of Mathematical Physics*, AMS Colloquium Publications 49 (Providence, RI: American Mathematical Society).

Christensen, J. P. R. (1973) Measure theoretic zero sets in infinite dimensional spaces and applications to differentiability of Lipschitz mappings. *Publ. Dép. Math. (Lyon)* **10**, 29–39.

Chueshov, I. D. (2002) *Introduction to the Theory of Infinite-dimensional Dissipative Systems* (Kharkiv, Ukraine: ACTA Scientific Publishing House).

Constantin, P. & Foias, C. (1985) Global Lyapunov exponents, Kaplan–Yorke formulas and the dimension of the attractor for 2D Navier–Stokes equation. *Commun. Pure. Appl. Math.* **38**, 1–27.

Constantin, P. & Foias, C. (1988) *Navier–Stokes Equations* (Chicago, IL: University of Chicago Press).

Constantin, P., Foias, C., & Temam, R. (1988) On the dimension of the attractors in two-dimensional turbulence, *Physica D* **30**, 284–296.

Constantin, P., Foias, C., Nicolaenko, B., & Temam, R. (1989) Spectral barriers and inertial manifolds for dissipative partial differential equations. *J. Dynam. Diff. Eq.* **1**, 45–73.

Davies, E. B. (1995) *Spectral Theory and Differential Operators*, Cambridge Studies in Advanced Mathematics 42 (Cambridge: Cambridge University Press).

Debussche, A. and Temam, R. (1994) Convergent families of approximate inertial manifolds. *J. Math. Pures Appl.* **73**, 489–522.

Doering, C. R. & Gibbon, J. D. (1995) *Applied Analysis of the Navier–Stokes Equations*, Cambridge Texts in Applied Mathematics (Cambridge: Cambridge University Press).

Douady, A. & Oesterlé, J. D. (1980) Dimension de Hausdorff des attracteurs. *C. R. Acad. Sci. Paris Sr. A–B* **290 A**, 1135–1138.

Eden, A., Foias, C., Nicolaenko, B., & Temam, R. (1994) *Exponential Attractors for Dissipative Evolution Equations*, Research in Applied Mathematics Series (New York, NY: Wiley).

Edgar, G. (2008) *Measure, Topology, and Fractal Geometry*, second edition, Undergraduate Texts in Mathematics (New York, NY: Springer).

Edgar, G. A. (1998) *Integral, Probability, and Fractal Measures* (New York, NY: Springer).

Engelking, R. (1978) *Dimension Theory* (Amsterdam: North-Holland).

Evans, L. C. (1998) *Partial Differential Equations*, Graduate Studies in Mathematics 19 (Providence, RI: American Mathematical Society).

Evans, L. C., & Gariepy, R. F. (1992) *Measure Theory and Fine Properties of Functions*, Studies in Advanced Mathematics (Boca Raton, FL: CRC Press).

Falconer, K. J. (1985) *The Geometry of Fractal Sets*, Cambridge Tracts in Mathematics (Cambridge: Cambridge University Press).

Falconer, K. J. (1990) *Fractal Geometry*, Mathematical foundations and applications (Chichester: Wiley).

Falconer, K. J. (1997) *Techniques in Fractal Geometry* (Chichester: Wiley).

Federer, H. (1969) *Geometric Measure Theory* (New York, NY: Springer).

Flores, A. (1935) Über n-dimensionale Komplexe die im R_{2n+1} absolut selbstverschlungen sind. *Ergebnisse eines mathematischen Kolloquiums* **6** (1933/1934), 4–7. Reprinted in Menger (1998).

Foias, C. & Olson, E. J. (1996) Finite fractal dimension and Hölder–Lipschitz parametrization. *Indiana Univ. Math. J.* **45**, 603–616.

Foias, C. & Temam, R. (1984) Determination of the solutions of the Navier–Stokes equations by a set of nodal values. *Math. Comp.* **43**, 117–133.

Foias, C. & Temam, R. (1989) Gevrey class regularity for the solutions of the Navier-Stokes equations. *J. Funct. Anal.* **87**, 359–369.

Foias, C. & Titi, E. S. (1991) Determining nodes, finite difference schemes and inertial manifolds. *Nonlinearity* **4**, 135–153.

Foias, C., Manley, O., & Temam, R. (1988) Modelling the interaction of small and large eddies in two-dimensional turbulent flows. *RAIRO Math. Model. Numer. Anal.* **22**, 93–114.

Friz, P. K. & Robinson, J. C. (1999) Smooth attractors have zero "thickness". *J. Math. Anal. Appl.* **240**, 37–46.

Friz, P. K. & Robinson, J. C. (2001) Parametrising the attractor of the two-dimensional Navier–Stokes equations with a finite number of nodal values. *Physica D* **148**, 201–220.

Friz, P. K., Kukavica, I., & Robinson, J. C. (2001) Nodal parametrisation of analytic attractors. *Disc. Cont. Dyn. Sys.* **7**, 643–657.

Gardner, R. J. (2002) The Brunn–Minkowski inequality. *Bull. Amer. Math. Soc.* **39**, 355–405.

Gilbarg, D. & Trudinger, N. S. (1983) *Elliptic Partial Differential Equations of Second Order*, Grundlehren der mathematischen Wissenschaften 224 (Berlin: Springer-Verlag).

Gromov, M. (1999) *Metric Structures for Riemannian and Non-Riemannian Spaces*, Progr. Math. vol. 152 (Boston, MA: Birkhäuser).

Guillopé, C. (1982) Comportement á l'infini des solutions des équations de Navier–Stokes et propriété des ensembles fonctionnels invariants (ou attracteurs). *C. R. Acad. Sci. Paris Ser. I Math.* **294**, 221–224.

Hale, J. K. (1988) *Asymptotic Behavior of Dissipative Systems*, Mathematical Surveys and Monographs 25 (Providence, RI: American Mathematical Society).

Hale, J. K., Magalhães, L. T., & Oliva, W. M. (2002) *Dynamics in Infinite Dimensions*, second edition, Applied Mathematical Sciences 47 (New York, NY: Springer).

Heinonen, J. (2001) *Lectures on Analysis on Metric Spaces*, Universitext (New York, NY: Springer-Verlag).

Heinonen, J. (2003) *Geometric Embeddings of Metric Spaces*, Report University of Jyväskylä Department of Mathematics and Statistics, 90 (Jyväskylä: University of Jyväskylä).

Henry, D. (1981) *Geometric Theory of Semilinear Parabolic Equations*, Springer Lecture Notes in Mathematics 840 (Berlin: Springer-Verlag).

Hensley, D. (1979) Slicing the cube in $\mathbb{R}^n$ and probability. *Proc. Amer. Math. Soc.* **73**, 95–100.

Heywood, J. G. & Rannacher, R. (1982) Finite approximation of the nonstationary Navier–Stokes problem. Part I: Regularity of solutions and second-order error estimates for spatial discretization. *SIAM J. Numer. Anal.* **19**, 275–311.

Howroyd, J. D. (1995) On dimension and on the existence of sets of finite positive Hausdorff measure. *Proc. London Math. Soc.* **70**, 581–604.

Howroyd, J. D. (1996) On Hausdorff and packing dimension of product spaces. *Math. Proc. Cambridge Philos. Soc.* **119**, 715–727.

Hunt, B. (1996) Maximal local Lyapunov dimension bounds the box dimension of chaotic attractors. *Nonlinearity* **9**, 845–852.

Hunt, B. R. & Kaloshin, V. Y. (1997) How projections affect the dimension spectrum of fractal measures. *Nonlinearity* **10**, 1031–1046.

Hunt, B. R. & Kaloshin, V. Y. (1999) Regularity of embeddings of infinite-dimensional fractal sets into finite-dimensional spaces. *Nonlinearity* **12**, 1263–1275.

Hunt, B. R., Sauer, T., & Yorke, J. A. (1992) Prevalence: a translation-invariant almost every for infinite dimensional spaces. *Bull. Amer. Math. Soc.* **27**, 217–238; (1993) Prevalence: an addendum. *Bull. Amer. Math. Soc.* **28**, 306–307.

Hurewicz, W & Wallman, H. (1941) *Dimension Theory*, Princeton Math. Ser. 4 (Princeton, NJ: Princeton University Press).

John, F. (1948) Extremum problems with inequalities as subsidiary conditions, pages 187–204 in *Studies and Essays Presented to R. Courant on his 60th Birthday, January 8, 1948* (New York, NY: Interscience Publishers, Inc., New York).

John, F. (1982) *Partial Differential Equations*, fourth edition (New York, NY: Springer).

Johnson, W. & Lindenstrauss, J. (1984) Extensions of Lipschitz maps into a Hilbert space. *Contemp. Math.* **26**, 189–206.

Kadec, M. I. & Snobar, M. G. (1971) Certain functionals on the Minkowski compactum. *Math. Notes* **10**, 694–696.

Kakutani, S. (1939) Some characterizations of Euclidean space. *Japanese J. Math.* **16**, 93–97.

Katětov, M. (1952) On the dimension of non-separable spaces. I. *Czech. Math. Journ.* **2**, 333–368.

Kraichnan, R. H. (1976) Inertial ranges in two-dimensional turbulence. *Phys. Fluids* **10**, 1417–1423.

Kukavica, I. (1994) An absence of a certain class of periodic solutions in the Navier–Stokes equations. *J. Dynam. Differential Equations* **6**, 175–183.

Kukavica, I. (2007) Log–log convexity and backward uniqueness. *Proc. Amer. Math. Soc.* **135**, 2415–2421.

Kukavica, I. & Robinson, J. C. (2004) Distinguishing smooth functions by a finite number of point values, and a version of the Takens embedding theorem. *Physica D* **196**, 45–66.

Laakso, T. J. (2002) Plane with A_∞-weighted metric not bi-Lipschitz embeddable to $\mathbb{R}^N$. *Bull. London Math. Soc.* **34**, 667–676.

Landau, L. D. & Lifshitz, E. M. (1959) *Fluid Mechanics*, Course of Theoretical Physics 6. (London: Pergamon, London).

Lang, U. & Plaut, C. (2001) Bilipschitz embeddings of metric spaces into space forms. *Geom. Dedicata* **87**, 285–307.

Langa, J. A. & Robinson, J. C. (2001) A finite number of point observations which determine a non-autonomous fluid flow. *Nonlinearity* **14**, 673–682.

Langa, J. A. & Robinson, J. C. (2006) Fractal dimension of a random invariant set. *J. Math. Pures Appl.* **85**, 269–294.

Larman, D. G. (1967) A new theory of dimension. *Proc. London Math. Soc.* **17**, 178–192.

Liu, V. X. (1993) A sharp lower bound for the Hausdorff dimension of the global attractors of the 2D Navier–Stokes equations. *Comm. Math. Phys.* **158**, 327–339.

Luukkainen, J. (1981) Approximating continuous maps of metric spaces into manifolds by embeddings. *Math. Scand.* **49**, 61–85.

Luukkainen, J. (1998) Assouad dimension: antifractal metrization, porous sets, and homogeneous measures. *J. Korean Math. Soc.* **35**, 23–76.

Mallet-Paret, J. (1976) Negatively invariant sets of compact maps and an extension of a theorem of Cartwright. *J. Differential Equations* **22**, 331–348.

Mandelbrot, B. B. (1982) *The Fractal Geometry of Nature* (W. H. Freeman & Co, San Francisco, CA).

Mañé, R. (1981) On the dimension of the compact invariant sets of certain nonlinear maps. In Rand, D. A., & Young, L. S. (eds) *Dynamical Systems and Turbulence,*

Warwick 1980, Springer Lecture Notes in Mathematics 898, pp. 230–242 (New York, NY: Springer).
Mattila, P. (1975) Hausdorff dimension, orthogonal projections and intersections with planes. *Ann. Acad. Sci. Fennicae A* **1**, 227–244.
Mattila, P. (1995) *Geometry of Sets and Measures in Euclidean Spaces*, Cambridge Studies in Advanced Mathematics (Cambridge: Cambridge University Press).
Meise, R. & Vogt, D. (1997) *Introduction to Functional Analysis* (Oxford: Oxford University Press).
Menger, K. (1926) Über umfassendste n-dimensionale Mengen. *Proc. Akad. Wetensch. Amst.* **29**, 1125–128.
Menger, K. (1998) *Ergebnisse eines Mathematischen Kolloquiums*. Reissued and edited by E. Dierker & K. Sigmund (Vienna: Springer).
Morita, K. (1954) Normal families an dimension theory for metric spaces. *Math. Ann.* **128**, 350–362.
Movahedi-Lankarani, H. (1992) On the inverse of Mañé's projection, *Proc. Amer. Math. Soc.* **116**, 555–560.
Munkres, J. R. (2000) *Topology*, second edition (Upper Saddle River, NJ: Prentice Hall).
Nöbeling, G. (1931) Über eine n-dimensionale Universalmenge im R^{2n+1}. *Math. Ann.* **104**, 71–80.
Olson, E. J. (2002) Bouligand dimension and almost Lipschitz embeddings. *Pacific J. Math.* **2**, 459–474.
Olson, E. J. & Robinson, J. C. (2010) Almost bi-Lipschitz embeddings and almost homogeneous sets. *Trans. Amer. Math. Soc.* **362**, 145–168.
Ott, W. & Yorke, J. A. (2005) Prevalence. *Bull. Amer. Math. Soc.* **42**, 263–290.
Ott, W., Hunt, B. R., & Kaloshin, V. Y. (2006) The effect of projections on fractal sets and measures in Banach spaces. *Ergodic Theory Dynam. Systems* **26**, 869–891.
Pesin, Ya. B. (1997) *Dimension Theory in Dynamical Systems: Contemporary Theory and Applications*, Chicago Lectures in Mathematics (Chicago, IL: University of Chicago Press).
Pinto de Moura, E. & Robinson, J. C. (2010a) Orthogonal sequences and regularity of embeddings into finite-dimensional spaces. *J. Math. Anal. Appl.* **368**, 254–262.
Pinto de Moura, E. & Robinson, J. C. (2010b) Lipschitz deviation and embeddings of global attractors. *Nonlinearity* **23**, 1695–1708.
Pinto de Moura, E. & Robinson, J. C. (2010c) Log-Lipschitz continuity of the vector field on the attractor of certain parabolic equations, arXiv preprint 1008.4949.
Pinto de Moura, E., Robinson, J. C., & Sánchez-Gabites, J. J. (2010) Embedding of global attractors and their dynamics, *Proc. Amer. Math. Soc.*, to appear.
Pontrjagin, L. & Schnirelmann, G. (1932) Sur une propriété métrique de la dimension. *Ann. Math.* **33**, 156–162.
Prosser, R. T. (1970) Note on metric dimension. *Proc. Amer. Math. Soc.* **25**, 763–765.
Robinson, C. (1995) *Dynamical Systems: Stability, Symbolic Dynamics, and Chaos* (London: CRC Press).
Robinson, J. C. (1997) Some closure results for inertial manifolds. *J. Dyn. Diff. Eq.* **9**, 373–400.
Robinson, J. C. (2001) *Infinite-dimensional Dynamical Systems*, Cambridge Texts in Applied Mathematics (Cambridge: Cambridge University Press).

Robinson, J. C. (2005) A topological delay embedding theorem for infinite-dimensional dynamical systems. *Nonlinearity* **18**, 2135–2143.

Robinson, J. C. (2007) Parametrization of global attractors, experimental observations, and turbulence. *J. Fluid Mech.* **578**, 495–507.

Robinson, J. C. (2008) A topological time-delay embedding theorem for infinite-dimensional cocycle dynamical systems. *Discrete Contin. Dyn. Syst. Ser. B* **9**, 731–741.

Robinson, J. C. (2009) Linear embeddings of finite-dimensional subsets of Banach spaces into Euclidean spaces. *Nonlinearity* **22**, 711–728.

Robinson, J. C. (2010) Log-Lipschitz embeddings of homogeneous sets with sharp logarithmic exponents and slicing the unit cube, arXiv preprint 1007.4570.

Robinson, J. C. & Vidal-López, A. (2006) Minimal periods of semilinear evolutions equations with Lipschitz nonlinearity. *J. Diff. Eq.* **220**, 396–406.

Rogers, C. A. (1998) *Hausdorff Measures*, reissue of 1998 edition with a foreword by K. Falconer (Cambridge: Cambridge University Press).

Roman, S. (2007) *Advanced Linear Algebra*, third edition, Graduate Texts in Mathematics (New York, NY: Springer).

Rosa, R. (1995) Approximate inertial manifolds of exponential order. *Discrete Contin. Dynam. Systems* **1**, 421–448.

Sato, T. (2009) An alternative proof of Berg and Nikolaev's characterization of CAT(0)-spaces via quadrilateral inequality. *Archiv Math.* **93**, 487–490.

Sauer, T., Yorke, J. A., & Casdagli M. (1991) Embedology. *J. Stat. Phys.* **71**, 529–547.

Scheffer, V. (1976) Turbulence and Hausdorff dimension, pages 174–183 in *Turbulence and Navier–Stokes Equations, Orsay 1975*, Springer Lecture Notes in Mathematics 565 (Berlin: Springer-Verlag).

Sell, G. R. & You, Y. (2002) *Dynamics of Evolutionary Equations*, Applied Mathematical Sciences, 143 (New York, NY: Springer-Verlag).

Semmes, S. (1996) On the nonexistence of bilipschitz parametrizations and geometric problems about A_∞-weights. *Rev. Mat. Iberoamericana* **12**, 337–410.

Sharples, N. (2010) Strict inequality in the box-counting dimension product formulas, arXiv preprint 1007.4222.

Sontag, E. D. (2002) For differential equations with r parameters, $2r + 1$ experiments are enough for identification. *J. Nonlin. Sci.* **12**, 553–583.

Takens, F. (1981) Detecting strange attractors in turbulence. In Rand, D. A. & Young, L. S. (eds.) *Dynamical Systems and Turbulence, Warwick 1980*, Springer Lecture Notes in Mathematics 898, pp. 366–381 (New York: Springer).

Temam, R. (1977) *Navier–Stokes Equations* (Amsterdam: North-Holland; reprinted by AMS Chelsea, 2001).

Temam, R. (1988) *Infinite-Dimensional Dynamical Systems in Mechanics and Physics*, Springer Applied Mathematical Sciences 68 (Berlin: Springer-Verlag).

Tricot, C. (1980) Rarefaction indices. *Mathematika* **27**, 46–57.

Wells, J. H. & Williams, L. R. (1975) *Embeddings and Extensions in Analysis* (New York, NY: Springer-Verlag).

Whitney, H. (1936) Differentiable manifolds. *Ann. Math.* **37**, 645–680.

Yorke, J. A. (1969) Periods of periodic solutions and the Lipschitz constant. *Proc. Am. Math. Soc.* **22**, 509–512.

Index

Printed in the United States
by Baker & Taylor Publisher Services